Engineering Materials

This series provides topical information on innovative, structural and functional materials and composites with applications in optical, electrical, mechanical, civil, aeronautical, medical, bio- and nano-engineering. The individual volumes are complete, comprehensive monographs covering the structure, properties, manufacturing process and applications of these materials. This multidisciplinary series is devoted to professionals, students and all those interested in the latest developments in the Materials Science field, that look for a carefully selected collection of high quality review articles on their respective field of expertise.

Indexed at Compendex (2021)

More information about this series at https://link.springer.com/bookseries/4288

Md Rezaur Rahman · Chin Mei Yun ·
Muhammad Khusairy Bin Bakri
Editors

Waste Materials in Advanced Sustainable Concrete

Reuse, Recovery and Recycle

Springer

Editors
Md Rezaur Rahman
Faculty of Engineering
Universiti Malaysia Sarawak
Kota Samarahan, Sarawak, Malaysia

Muhammad Khusairy Bin Bakri
Faculty of Engineering
Universiti Malaysia Sarawak
Kota Samarhan, Sarawak, Malaysia

Chin Mei Yun
Faculty of Engineering Computing and Science
Swinburne University of Technology
Kuching, Sarawak, Malaysia

ISSN 1612-1317 ISSN 1868-1212 (electronic)
Engineering Materials
ISBN 978-3-030-98811-1 ISBN 978-3-030-98812-8 (eBook)
https://doi.org/10.1007/978-3-030-98812-8

This Springer imprint is published by the registered company Springer Nature Switzerland AG
The registered company address is: Gewerbestrasse 11, 6330 Cham, Switzerland

Contents

Pervious Concrete Properties and Its Applications

Chin Mei Yun, Md Rezaur Rahman, Kuok King Kuok, Amelia Chai Pei Sze, Adrian Lee Kah Seng, and Muhammad Khusairy Bin Bakri

Abstract The purpose of this study is to investigate the different characteristics of Portland Cement Pervious (PCP). It has the potential to address a few critical environmental concerns, including groundwater recharging and stormwater runoff reduction. The lack of fine aggregate in the concrete mix results in pervious concrete interconnected empty spaces in the hardened matrix. Compared to conventional Portland Cement (CPC) concrete, these linked void spaces allow the concrete to transport water at comparatively high rates. Pervious concrete research has been limited due to differences in aggregate size and mix ratio, which impact the mechanical and hydrological characteristics of pervious concrete. As a result, this study looks at the best aggregate size and mix ratio for pervious concrete without fine aggregates. This study aims to determine the amount of coarse aggregate, cement, and water needed in the design mix ratio, considering the workability, compressive strength, and permeability as criteria. Samples were characterized using a scanning electron microscope (SEM), energy dispersive spectroscopy (EDS/EDX) Fourier transform infrared spectroscopy (FTIR), which will be used to support the data and analysis. The experimental study was carried out by collecting all essential data and manipulating the size of aggregate used in concrete samples. The optimal value of key parameters such as the water to cement ratio and the amount of aggregate to binder ratio was determined through the trial tests.

Keywords Compressive · Strength · Permeability · Coarse · Aggregate

C. M. Yun · K. K. Kuok · A. C. P. Sze · A. L. K. Seng
Faculty of Engineering, Computing and Science, Swinburne University of Technology Sarawak Campus, Jalan Simpang Tiga, 93400 Kuching, Sarawak, Malaysia

M. R. Rahman (✉) · M. K. B. Bakri
Faculty of Engineering, Universiti Malaysia Sarawak, Jalan Datuk Mohammad Musa, 94300 Kota Samarahan, Sarawak, Malaysia
e-mail: rmrezaur@unimas.my

M. R. Rahman et al. (eds.), *Waste Materials in Advanced Sustainable Concrete*,
Engineering Materials, https://doi.org/10.1007/978-3-030-98812-8_1

1 Introduction to Pervious Concrete

Pervious concrete is a composite material made of Portland cement, coarse aggregate, and water that does not contain fine particles (Ibrahim et al., 2014). Pervious concrete, in other words, is constructed using large coarse aggregates particle size with little to no fine aggregates. Porous concrete is also known as permeable concrete, no-fines concrete, and porous pavement. The large coarse aggregates are then coated with cement paste, creating high-porosity concrete used for concrete pavement, allowing water from precipitation and other sources to pass through directly. Thus, minimizing runoff and facilitating groundwater recharge (Obla, 2010). Due to its large porosity, prior concrete has a lower compressive strength than ordinary concrete (Alsayed & Amjad, 1996; Kim et al., 2014). As a result, the restricted usage of prior concrete in particular applications should not be seen as a replacement for conventional concrete in real-world buildings (Gardner et al., 2018).

Pervious concrete has a greater cement content, lower compressive loadings, more porosity, and a lower unit weight than ordinary concrete (Hariyadi & Tamai, 2015; Kovac & Sicakova, 2017). Pervious concrete is commonly utilized in parking lots, low-traffic areas, residential streets, pedestrian walkways, and greenhouse applications (ACI Committee 522, 2010). It is also an important application for sustainable construction and is one of many low impact development techniques used by builders to protect water quality.

People are becoming more aware of the benefits of pervious concrete, such as sound isolation of road pavement, environmental protection, and stormwater reduction, because of current development and increasing demand for pervious concrete on recent sustainable construction activities (Barisic et al., 2017; Dash & Kar, 2018; Gupta, 2014; Hamdy, 2016; Kovac & Sicakova, 2017). However, pervious concrete could only be used for modest pavement construction (Kia et al., 2021a, 2021b; Yao et al., 2018; Yu et al., 2019). As a result, more laboratory tests should be conducted to obtain compressive strength and porosity information.

Pervious concrete was originally utilized as pavement surface and load-bearing walls in Europe in the 1800s (Dash & Kar, 2018; Wanielista et al. 2007). As the amount of cement used was reduced, cost efficiency became the primary motivation. In Scotland and England, it gained popularity again in the 1920s for two-story residences. Due to the scarcity of cement in Europe after WWII, it became increasingly feasible. As for the United States, It was not until the 1970s that it gained popularity in the United States (Wanielista et al., 2007). In 2000, pervious concrete became well-known in India, Malaysia, and other Asian nations for its innovative and successful approach to addressing critical environmental challenges and promoting green, sustainable growth. Porous concrete plays an important role in recharging groundwater, decreasing stormwater runoff, and complying with environmental protection agency (EPA) stormwater rules by collecting stormwater and enabling it to sink into the earth (Kubba, 2010; Lee et al., 2012).

Impervious pavement covers up to 75% of the urban surface area, preventing groundwater recharge, contributing to floods and erosion, transporting pollutants to nearby waterways, and increasing the complexity and cost of stormwater treatment (Ma et al., 2021; McGrane, 2016). As opposed to the impervious pavement, one of the most distinguishing features of pervious pavement is the presence of gaps that enable water to seep through to the foundation materials beneath, which improves groundwater recharge while reducing peak stormwater flow and pollutants (Pilon et al., 2018). Recycled aggregate and fly ash can be used in pervious pavement, reducing waste and embodied energy (Kubba, 2010). Pervious pavement with a compressive strength of up to 4000 psi is appropriate for use in parking and access areas (Kubba, 2010). It also reduces tree root issues, and the percolation area stimulates deeper root growth (Kubba, 2010). Due to improved heat exchange with the underlying soil, summer ambient air temperature can be reduced by 2–4 °F thanks to enhanced heat exchange with the underlying soil (Kubba, 2010).

According to studies on both mechanical and hydrological properties of pervious concrete, it can be inferred that lowering the size of aggregate used would improve mechanical features of pervious concrete, such as compressive strength. The size of aggregate used might be increased to improve its hydrological qualities, such as permeability. When performing an inverted slump test for a concrete mixture, the workability of the concrete is satisfactory if no cement paste leaks beneath the slump cone (Jimma, 2014). To achieve consistent pore sizes in PCP concrete, it is also necessary to use uniform aggregate size. The aggregate size used in PCP concrete typically varies from 9.5 to 19.5 mm (Sonebi et al., 2016). Larger aggregate sizes may reduce compressive strength since there is less contact area between the aggregates, resulting in lower binder strength. However, because of the greater pore size, the permeability of PCP concrete might be improved by using bigger size particles. As the water-to-cement ratio might impact the compressive strength of PCP concrete, the best water-to-cement ratio is 0.33, according to many water-to-cement ratios used by different studies.

2 Existing Pervious Concrete and Pervious Geopolymer Concrete

Concrete with a high void content that allows water or air to infiltrate is known as pervious concrete, porous concrete, or water-permeable concrete. In general, pervious concrete contains pores ranging from 2 to 8 mm, 18 to 35% void content, and compressive strength of 2.8–28.0 MPa (ACI Committee 522, 2010). In addition to Portland cement, Geo-polymers can be utilized as cementitious ingredients in the production of pervious concrete. The characteristics of pervious geopolymers are comparable to those of pervious concrete made of Portland cement (Jang et al., 2015; Sata et al., 2013; Tho-in et al., 2012; Zaetang et al., 2015).

Both Portland cement and geopolymer concrete can use construction waste materials such as old concrete, block concrete, or clay brick aggregates. They can also be used to replace natural aggregates in pervious geopolymer concrete. Pervious concretes made from recycled aggregates are compared to natural aggregates in terms of surface roughness (Sata & Chindaprasirt, 2020). The densities of pervious geopolymer concrete are lower than those of ordinary concrete due to the large void fraction, with values ranging from 1420 to 1840 kg/m^3 (Sata & Chindaprasirt, 2020). The use of sodium hydroxide solution at various concentrations (10, 15, and 20 M) appears to have a minor effect on concrete density (Sata & Chindaprasirt, 2020). The density of coarse aggregates from reinforced concrete block (RCB) was the lowest, at 1420 kg/m^3. In contrast, limestone, recycled concrete, and recycled block concrete (RBC) aggregates had averages of 1800, 1720, and 1650 kg/m^3, respectively (Sata & Chindaprasirt, 2020). The decreased dry-rodded density of RCB caused this (Wongsa, 2013).

The compressive strength of crushed limestone pervious geopolymer concrete (L.S.) ranges between 11.9 and 13.6 MPa. The compressive strengths of recycled concrete (RCA), reinforced brick concrete (RBC), and RCB pervious geopolymer concrete were lower, ranging from 7.0 to 10.3 MPa, 2.8–3.8 MPa, and 2.9–6.6 MPa, respectively (Sata & Chindaprasirt, 2020). According to ACI Committee 522 (2010), these compressive strengths were within the range of ordinary pervious concrete. The effect of adding recycled aggregates on tensile strength was comparable to the impact on compressive strength. According to test findings, the water permeability of pervious geopolymer concrete is measured in the range of 0.71–1.80 m/s. The void content of the RCB pervious geopolymer concrete appears to be lower than when crushed limestone, recycled concrete, and RBC was used. Pervious concrete's water permeability coefficient followed the same trend as the void content. When RCB was employed, the water permeability coefficient was the lowest, with values ranging from 0.71 to 1.2 cm/s. Due to the poor strength of clay bricks, their particles were shattered during concrete mixing, resulting in pervious geopolymer concrete with low water permeability and void content (Wongsa, 2013).

Bottom ash can be used as coarse particles in pervious geopolymer concrete when a lower density than regular pervious geopolymer concrete, with density values ranging from 1470 to 1500 kg/m^3. The compressive strength is 5.9–8.6 MPa, the thermal conductivity is 0.30–0.33 W/mK, and the ultrasonic pulse velocity is 1726–2566 m/s, which is quite low compared to common pervious concrete (Zaetang et al., 2015). Low density, compressive strength, thermal conductivity, and ultrasonic pulse velocity define the bottom ash particles with a significant void content and a rough surface. Furthermore, heavy metals can be successfully immobilized as solidified products using pervious concrete made from geopolymer binders and bottom ash particles (Jang et al., 2015).

3 Pervious Concrete Engineering Properties

Pervious concrete has a lower compressive strength, improved permeability, and a lower unit weight than regular concrete, about 70% of conventional concrete.

3.1 Density

In comparison to traditional concrete, which has a void ratio of around 3–5%, pervious concrete has a void ratio of 15–40%, depending on the application. The large vacancy ratio results in a lower unit weight of roughly 70% of ordinary concrete (Ajamu et al., 2012).

3.2 Porosity/Permeability

The sample porosity is defined by the volume of voids divided by the total volume of the sample. A water displacement method can determine the overall porosity of pervious concrete (Montes et al., 2005). Porosity should be in the range of 15–25% (Tennis et al., 2004). The porosity of a concrete mix is determined by the water-to-cement ratio as well as the compaction effort (ACI Committee 522, 2010). Greater infiltration rates are possible with higher porosities, also known as void contents and void ratios, although compressive strength is considerably reduced. Although various methods have been tried, a falling-head device developed from soil testing is used to determine the permeability of porous concrete. According to the findings, permeability rose exponentially as the void ratio increased, and porous concrete with a porosity of less than 15% had limited or no permeability (Montes et al., 2005).

3.3 Compressive Strength

Pervious concrete's compressive strength can range from 3.45 to 27.58 MPa, whereas conventional concrete's compressive strength typically varies from 20.68 to 48.26 MPa (Hamdy, 2016). Flexural strengths of pervious concrete and traditional concrete usually differ from 1 to 3.8 MPa and 4 to 5.5 MPa, respectively (Tennis et al., 2004). Although porosity is the most important factor in compressive and flexural strength, aggregate size, shape, and gradation can impact pervious concrete strength (Crouch et al., 2007). The compressive strength of porous concrete is directly proportional to the unit weight of the mix, and the unit weight grows along with the strength. According to certain research, with the right water-cement mix and densification procedure, strengths of around 21 MPa or higher are easily achievable (Schaefer et al., 2006). Using admixtures, however, pervious concrete strengths are higher than conventional and traditional concrete strengths, which may be achieved, reaching up to 55.16 MPa). It has been demonstrated that when the compaction energy or densification effort on the sample rises, the sample's compressive strength increases as well. This phenomenon is related to reducing air spaces, which may result in a considerable reduction in permeability. Because the major aim of a porous pavement system is to obtain enough permeability for stormwater control, compacting concrete until it reaches adequate strength is not always a possibility, and a compromise must be struck between strength and void ratio (Schaefer et al., 2006).

3.4 Durability

Pervious concrete's surface durability was already an issue. Even with adequate batching, handling, and curing, ravelling can occur in pervious concrete. The surface durability of pervious concrete may be tested using existing ASTM test methods (ASTM C944/C944M-19, 2019; ASTM C1747/C1747M-13, 2013). A laboratory test for determining the surface durability of pervious concrete has been suggested, which measures the ravelling of pervious concrete (Offenberg, 2006). Distress surveys were conducted on two separate field locations as part of research into structural performance, and the data were utilized to generate a pavement condition index (PCI). The high PCI ratings of the thickest pervious concrete sections revealed that, when correctly planned, pervious concrete may be utilized for most "Residential" and "Collector" roadways for conventional design life periods while displaying adequate structural performance (Goede & Haselbach, 2011).

4 Methodology

4.1 Materials

Materials required for this experiment involve coarse aggregates, sand, water and cement. Cahya Mata Sarawak (CMS) Berhad, Sarawak, Malaysia, supplied the coarse aggregates, crushed limestone with average diameters ranging from 5 to 10 mm, 10 to 20 mm and 20 to 25 mm. It was adopted as the main component of mixtures and was kept in the lab at room temperature. Portable Water was used, but it was held in an empty pail at room temperature before mixing. Portland cement grade CEM I 42.5 N was manufactured by Cahya Mata Sarawak (CMS) Berhad, Sarawak, Malaysia, which complied with BS EN 197-1 (2014) standard. River sand has been used as part of the concrete mixture for this experiment. Particle Size Distribution (PSD) was adopted as a classification method for sand based on BS 882 (1992) standard.

4.2 Sample Preparation

In this experiment, there were two components of the tests that were completed. The permeability test was the first component of the investigation, and the compressive strength test was the second component. This portion of the research focused on a list of procedures that have been carried out in the preparation, mixing, casting, curing, and testing of PCP concrete. Particle Size Distribution (PSD) tests were carried out for the sand used in CPC concrete during the material preparation stage to identify distinct particle sizes based on BS 882 (1992). Trial mixes of each type of concrete were conducted to confirm that the mix ratio was adequate. A total of 54 samples were cast. CPC concrete is used in 27 control samples, whereas PCP concrete is used in 27 samples with aggregate sizes of 5–10 mm, 10–20 mm, and 20–25 mm. All coarse aggregates used for PCP concrete casting were sieved to ensure consistent size. Permeability tests were performed on all PCP samples after they were de-moulded and cured in the water tank for 7-days, 14-days, or 28-days. The slump, compressive strength, and permeability tests were all part of the three primary tests.

4.3 Method

4.3.1 Particle Size Distribution

Particle size distribution was utilized to categorize river sand particles into sizes that comply with BS 882 (1992) standard since it comprises a mixture of irregular shapes and sizes. Hence, sieve analysis with sieve sizes ranging from 5 to 75 m was used to identify the particle size distribution for river sand quantitatively. The particle size distribution of river sand falls within the top and lower grading limits in Fig. 1. This river sand is appropriate for mixing CPC concrete in this sample preparation.

4.3.2 Mix Proportions Determination and Concrete Mixing Preparation

Once all necessary materials have been prepared, do a trial mix using an initial mix proportion to acquire an optimal mix ratio. In Table 1, the mass of concrete components for each cube sample was determined and recorded. The prepared Portland cement and sand were first put into the drum mixer for the premixing procedure. Water was added to the concrete mix after one minute of premixing. Concrete mixtures were poured into the steel mould and allowed to cure for 24 h before being de-moulded. All concrete cube samples were immersed in the water tank in the lab for 7, 14, and 28 days of water curing after 24 h of hardening.

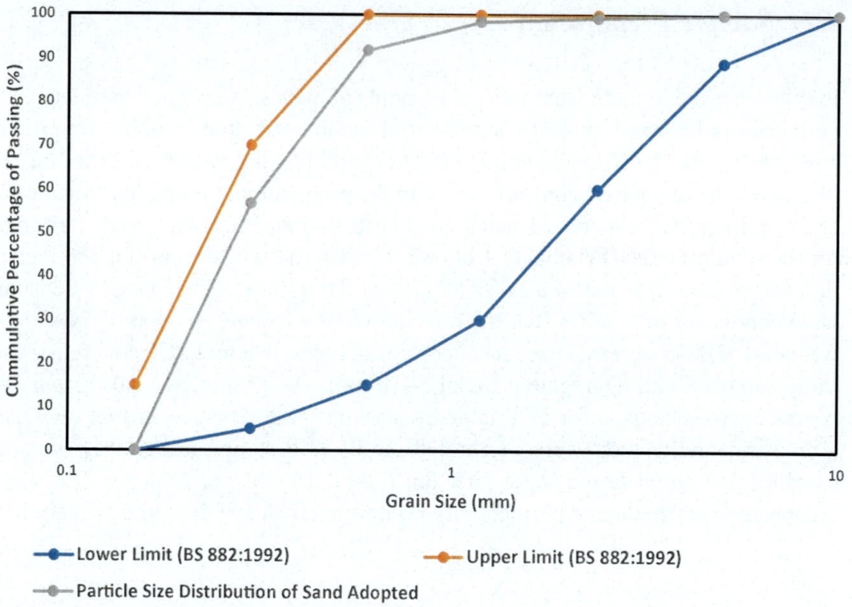

Fig. 1 Grading curve of sand particle

Table 1 Mass of concrete is required for each batch

Sample	CPC	PCP
Mix ratio	1:1:1.5	1:3
Water-cement ratio	0.38	0.38
Cement (kg/m^3)	736.84	736.84
Sand (kg/m^3)	736.84	-
Coarse aggregate (kg/m^3)	1105.26	2210.52
Water (kg/m^3)	280	280

4.4 *Test and Characterization*

4.4.1 Inverted Slump Test

Before casting, the inverted slump test was performed after the mixing procedure to ensure that each concrete batch was workable. ASTM C1611/C1611M-21 (2021) standard was used as a reference.

4.4.2 Compressive Strength Test

A compression strength test was performed to understand how would the concrete behave under compression stress. The compressive strength tests were performed using Unit Test Autocon 2000 under BS EN 12,390-3 (2019). During the compressive strength test, the load was increased progressively at 2.4 ± 0.2 kN/s across an area of 100 mm × 100 mm until the samples failed. The cube samples were loaded to failure on the 7th, 14th, and 28th days in compressive strength. The compressive strength of all cube samples was measured, and the results were presented.

4.4.3 Permeability Test

Permeability tests were performed on all PCP concrete samples to analyze the water infiltration characteristics of the beam. The permeability test was carried out in line with ASTM C1701/C1710M-17 (2017) standards. The PCP concrete samples were sealed to the testing equipment and allowed water to flow out of it. The amount of time it takes for water to flow out of the testing equipment has been recorded and graphed.

4.4.4 Scanning Electron Microscopy (SEM) and Energy Dispersive Spectroscopy (EDS/EDX)

Scanning electron microscopy (SEM) and energy dispersive spectroscopy (EDX/EDS) were performed under ASTM C1723-16 (2016) and ASTM E1508-12 (2019) standards, respectively. The samples were examined at magnifications of 100×, 250×, 500×, and 10000×. The elemental composition percentages of the samples are scanned and analyzed by automated software. The EDX/EDS was performed many times on each sample at various stages, with the most representative results chosen. For the materials' SEM and EDX/EDS examinations, Hitachi Ltd., Tokyo, Japan, utilized a Hitachi TM4000Plus Tabletop Microscope with a Quantax75TM Series Energy Dispersive X-Ray Spectrometer (Hitachi Ltd., Tokyo, Japan).

4.4.5 Fourier Transform Infrared Spectroscopy

FTIR materials were analyzed using Fourier-transform infrared spectroscopy (IRAffinity-1, Shimadzu Corporation, Kyoto, Japan). Fourier-transform infrared spectroscopy was performed for qualitative and quantitative examination according to ASTM E168-16 (2016) and ASTM E1252-98 (2021) standards. Each sample had its spectra scanned in the wavenumber range of 4000–400 cm^{-1}. The samples' infrared spectrum transmittance and absorption were used to create a unique chemical fingerprint spectrum using Fourier-transform infrared spectroscopy. For each sample, the test was performed many times, and the most representative findings were chosen.

5 Results and Discussions

5.1 *Inverted Slump Properties*

The workability of CPC and PCP concrete mixtures cannot be compared directly, as PCP concrete mixtures compatibility was tested using different testing methodologies. However, the water to binder ratio for both concrete mixes was set at 0.38 to keep the variables consistent. After two minutes, the slump cone filled with PCP concrete mixture, as shown in Fig. 2, shows no excessive cement paste leaking on the base plate. Therefore, the concrete mix has attained the proper water-to-cement ratio (Chan et al., 2018). If there is cement paste leakage at the base plate of the slump cone, the water to cement ratio of the concrete mixture used is likely to be too high (Beckemeier, 2012). Some researchers stated that the workability of the concrete mix could not be determined merely by looking at the base plate of the slump cone (Abd Elaty & Ghazy, 2016). Following the discovery of cement paste leakage, the slump cone was gently elevated to enable the PCP concrete mixture to flow out of the slump cone. Figures 3 and 4 shows how the concrete mix flowed easily out of the slump cone without clinging to the side walls. The PCP concrete combination can be regarded as a very workable concrete mixture in this situation.

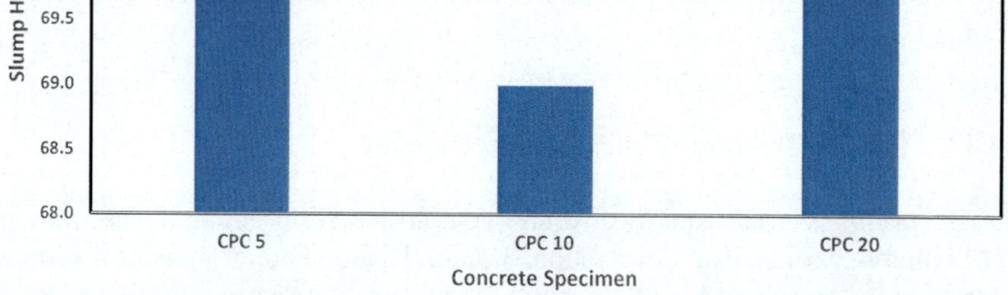

Fig. 2 Slump result of CPC concrete

Fig. 3 Evaluation of workability of inverted slump cone after two minutes

Fig. 4 Smooth flow of concrete out of slump cone

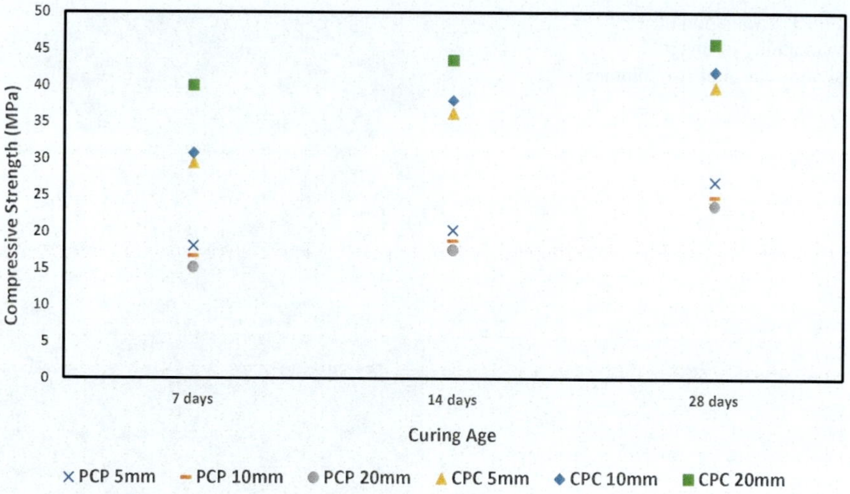

Fig. 5 Compressive strength results for PCP and CPC at different curing ages

5.2 *Compressive Strength Properties*

According to Fig. 5, the average compressive strength of PCP concrete rose as the aggregate size dropped. The compressive strength of CPC concrete, on the other hand, has grown as aggregate size has increased. The average compressive strength of PCP 20 mm after 7-days of curing time is 15.12 MPa, the lowest of all the concrete samples. Although PCP 5 mm reached the highest compressive strength of 18.10 MPa after just 7-days of curing, it could not surpass CPC concrete's minimum compressive strength of 29.41 MPa. Because the hydration process in concrete takes at least 28 days to complete, the compressive strength of all concrete samples is generally lowest after 7-days of curing. However, after the 14th day of curing age, the compressive strength of both PCP and CPC concrete samples began to grow. In 7 days of the curing period, the compressive strength of PCP 5 mm had increased from 18.10 to 20.32 MPa. This finding demonstrated that PCP 5 mm might be used as a building material for small road pavement, such as pedestrian walkways (Spandana & Rajasekhar, 2016). Because of its poor compressive strength, PCP concrete could not replace current CPC concrete for road pavement construction.

5.3 *Permeability Properties*

According to Fig. 6, PCP 5 has the lowest permeability with an aggregate size of 5-10 mm and an average permeability of 5.14 mm/s. It can be shown that the average permeability of both PCP 10 and PCP 20 with aggregate sizes of 10–20 mm and 20–25 mm has increased to 7.5 and 10.18 mm/s, respectively. Based on these findings,

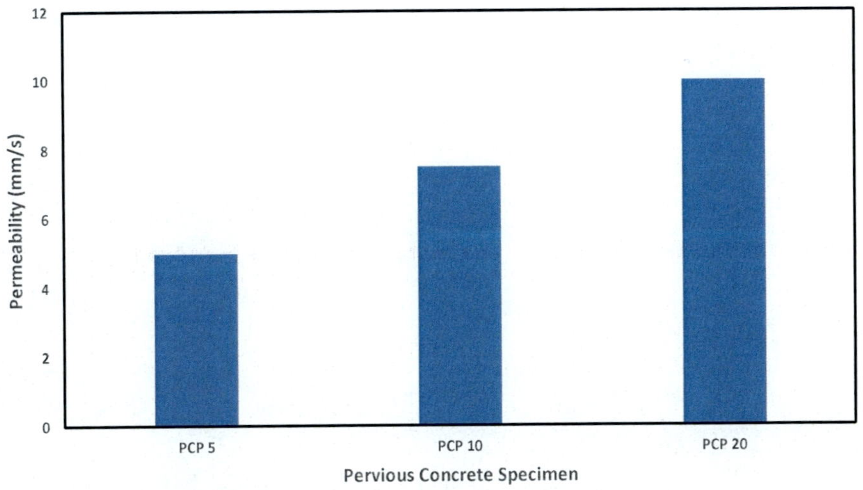

Fig. 6 Permeability for PCP concrete

it can be deduced that the larger the aggregate size of the sample, the greater the permeability of the concrete sample. As a result, PCP 10 is the best aggregate size for pedestrian sidewalk construction because it has appropriate compressive strength and permeability compared to other samples. Other researchers also reported a similar result (Cui et al., 2016; Huang et al., 2020; Kia et al., 2021a, 2021b; Qin et al., 2015; Vinay et al., 2020; Zhu et al., 2020).

5.4 Morphological Properties

Figure 7a–d shows the SEM images for PCP5 (28 days) at magnification of 100×, 250×, 500×, and 1000×, respectively. From Fig. 7, it is observed that the surface sample had a structured mini-rough hard-brittle topology. This surface allowed water to pass through perfectly while creating a smooth grip surface, preventing slipping. Hydraulic reactivity, or chemical or physical interaction, was used to generate this interlock with tiny particles when combined within themselves or with other materials (Rahman et al., 2017). The appropriate paste combination, soundness, and hydraulic reactivity were responsible for the compact microstructure (Chin et al., 2020a, 2020b; Henrist et al., 2003; Kumar et al., 2009; Shubbar et al., 2020; Wu et al., 2008). After a longer curing period, the microstructure became highly thick and coherent. Therefore, this indicates that the cementitious bond virtually completely covered the sample with fewer pore spaces (Chin et al., 2020a, 2020b, 2021; Henrist et al., 2003; Kumar et al., 2009; Shubbar et al., 2020; Wu et al., 2008). It can be concluded from the pictures in Fig. 7 that the correct mixing and chemical reactivity is necessary to

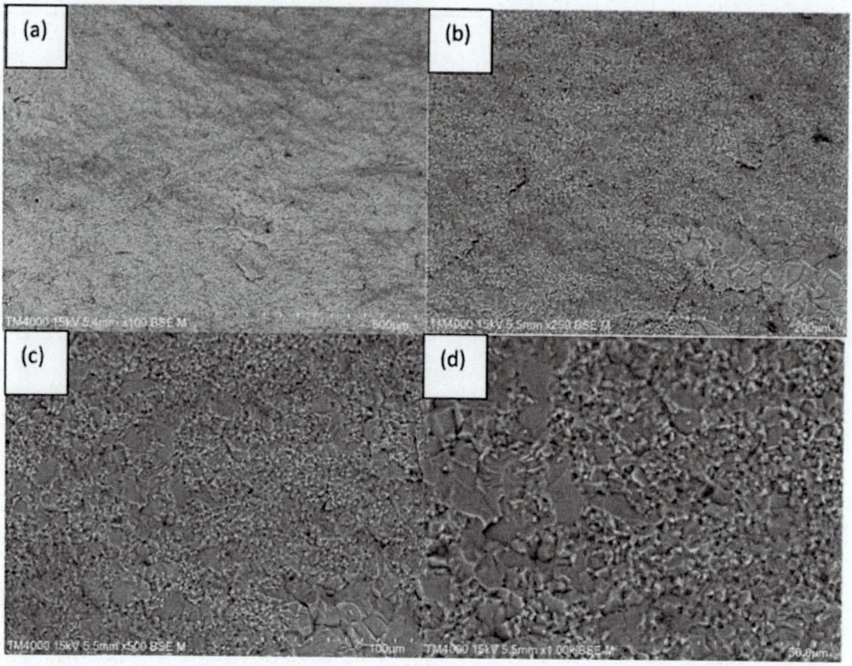

Fig. 7 SEM images for PCP5 (28 days) at magnification of **a** 100×, **b** 250×, **c** 500×, and **d** 1000×

form a good connection between each component utilized in the sample production, which helps enhance the compressive strength.

5.5 *Elemental Analysis*

Table 2 and Fig. 8 shows the EDS/EDX plot and element contents for the pervious concrete surface of PCP5 (28 days). The chemical elements present in the sample, according to Table 2, were carbon (C), oxygen (O), and calcium (Ca). Table 2 shows that oxygen mass percentage has the highest content, followed by calcium and carbon. Table 2 shows that chemical reactions happened during the paste mixture and hydraulic reactivity, which create calcium carbonate, $CaCO_3$ (Matschei et al.,

Table 2 Element composition for PCP5 (28 days)

Element	Mass normal (%)	Atom (%)
O	44.41	56.22
Ca	42.31	21.38
C	13.28	22.40
Sum	100.00	100.00

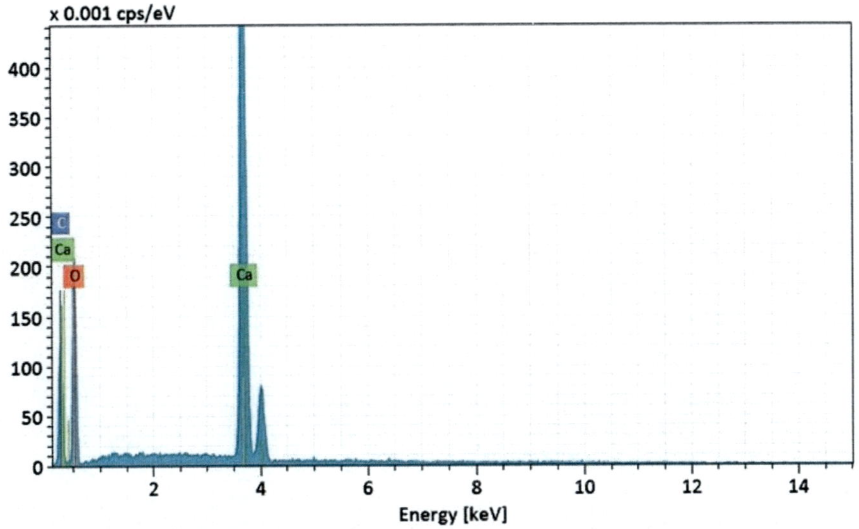

Fig. 8 EDS/EDX graph for PCP5 (28 days)

2007; Yaacob et al., 2015). The presence of $CaCO_3$ helps increase the early strength due to the accelerator effect and high rate of hydration, which hardens the concrete quicker (Matschei et al., 2007; Yaacob et al., 2015). At a mature age, the concrete with the $CaCO_3$ addition exhibits lower strength than concrete without $CaCO_3$, but still within the target strength (Matschei et al., 2007; Yaacob et al., 2015). It is also noted that proper mixing ratio and chemical substance used to create adequate reactivity creates a strong bonding between each substance used in the mixture for the sample, as shown through the morphological image by SEM.

5.6 *Infrared Spectral Properties*

Figure 9 shows the FTIR images for the pervious concrete surface of PCP5 (28 days). The small broad peak band at 3200–3500 cm^{-1} was due to the O–H bond's hydrophobic water bond and moisture. An analysis of the peak at 2216, 1701.22, and 1523.76 cm^{-1} was due to various FTIR spectra related to C-H bending and stretching. The most common wavelengths are 661.58 and 511.14 cm^{-1} due to symmetric and asymmetric valence and deformation vibration, respectively (Vaiciukyniene et al., 2013).

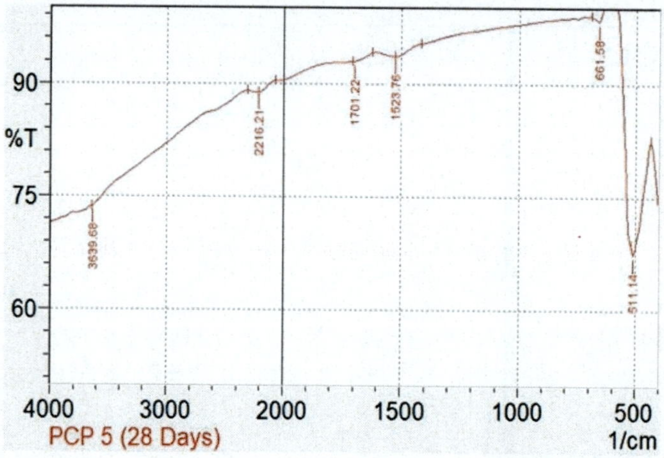

Fig. 9 FTIR for PCP5 (28 days)

5.7 *Applications, Maintenance, and Cost of Pervious Concrete*

Pervious concrete is important in aids of stormwater runoff (Alimohammadi et al., 2021). Pervious concrete is also the most effective stormwater management and treatment if properly designed (Alimohammadi et al., 2021). Pavement construction is the most common use for pervious concrete (Joshi & Dave, 2021). The U.S. Environmental Protection Agency (EPA) has even acknowledged this concrete as a solution that may provide the finest first-flush pollution reduction and stormwater management (Delatte & Schwartz, 2010). Drainage systems take up a lot of room and are also expensive. Pervious concrete holds water on-site, minimizing the size of storm sewers (Sartipi & Sartipi, 2019). Water is allowed to soak into the soil, resulting in a reduction in runoff. This concrete functions as an infiltration basin, allowing water to permeate and increasing groundwater recharge (Bonneau et al., 2017; Voisin et al., 2018). Thus, the diminishing levels of groundwater may be managed. Another benefit of pervious concrete pavement for stormwater management is that it filters rainwater before entering the system (Thives et al., 2018). As a result, the pollution load is kept out of the groundwater. This concrete also enables air and water to move through it. In addition, this contributes to the growth of trees since they too need enough air and water. It may therefore be beneficial to lay a strip of pervious concrete beside the pavement.

Because the pervious concrete is light in colour, it helps to save energy. Because the concrete reflects light, the demand for nighttime illumination is decreased (Li et al., 2013). According to the U.S. EPA, the first 1.5 inches of rainfall carry around 90% surface contaminants. Traditionally, stormwater drains do not redirect this water to treatment facilities instead of dumping it into local bodies of water. Consequently, algae development in the water body is accelerated, and aquatic life may suffer as a result. Its treatment costs a lot of money to make the water drinkable again. There are three primary causes of pollution that are commonly acknowledged. Sediment is one contaminant that is taken away in runoff. The elements of sediment are dirt and residues. Heavy metals are another contaminant, especially as they enter the discharge via the brake linings of automobiles. The hydrocarbons produce the third pollutant. The oil spilling from cars onto the pavement is one source of hydrocarbon (Hilpert et al., 2015). Asphalt, on the other hand, is a key source. According to research, the binder and sealer used in asphalt pavements contribute 90–95 percent of the hydrocarbon in urban runoff (Rordiguez-Hernandez et al., 2015). These are major environmental challenges that the EPA, regional watershed agencies, and municipal governments address by tightening stormwater management regulations. As a result, pervious concrete is becoming a preferred solution to this problem. The following are the reasons behind this: (i) It reduces the amount of untreated runoff water that reaches storm sewers by enhancing infiltration at the site; (ii) Aquifer levels are refilled as groundwater recharge is facilitated directly; (iii) More water is channelled towards trees and landscaping, reducing the need for irrigation water; (iv) Pollutants are prevented from contaminating watersheds, and thus the sensitive ecosystems are not harmed; and (v) Hydrocarbon pollution originating from asphalt pavements is also eliminated.

Its insulating properties complement its use as exterior load-bearing walls in single-story and multi-story structures. It provides heat insulation because of its reduced coefficient of thermal expansion. Because of its rough texture, plastering is also simple on pervious concrete walls. Pervious concrete with low capillary action resists moisture within the structure even when the exterior walls are wet. It is also appropriate in locations where fine sand is scarce. The use of pervious concrete reduces the urban heat island effect (Vujovic et al., 2021). Because the concrete is light in colour and has an open-cell structure, this is the case. As a result, pervious concrete absorbs less heat and radiates less heat into the environment (Zhang et al., 2015). Because of the colder temperature of the soil below, the pervious concrete's open void structure also helps keep the surface cool.

Pervious concrete offers many applications, including the parking lot and park pavements (Kia et al., 2017). The stagnant water puddles decrease when the previous concrete absorbs the water, making it easier to wander through parks and assuring the safety of cars on the road. The permeability of pervious concrete must be maintained for it to work properly. Cleaning should be done regularly to achieve this. The concrete surface is moist, and then vacuum sweeping is used to clean the surface. As sand is removed from the mix, the cost of pervious concrete drops dramatically. The cement used to make lean mixes is as low as 70–130 kg/m^3 of concrete. The price is greatly decreased because of the reduced usage of cement (Shinde & Valunjkar, 2015). In the United States, pervious concrete costs between \$2 and \$7 per square foot.

6 Conclusion

Finally, it was worth noting that the workability of CPC and PCP concrete improves as the water-to-cement ratio rises. For both types of concrete, the best water-to-cement ratio was 0.38. The compressive strength of PCP concrete diminishes when the aggregate size is increased. Compressive strength alone could not be used to determine the optimal aggregate size. As for the 14 days of curing time, PCP concrete with aggregate sizes of 5–10 mm has a compressive strength of 20.32 MPa. After 28 days of curing, these findings had exceeded the minimum compressive strength of 20.68 MPa. PCP concrete with aggregate sizes ranging from 5 to 10 mm can be used as a pedestrian sidewalk building material. PCP concrete with aggregate sizes of 20–25 mm had the greatest average water permeability of 10.18 mm/s, which is within the usual pervious concrete permeability range of 2–12 mm/s. The permeability of PCP concrete improves as the aggregate size increases, yet the compressive strength of the concrete decreases. The greatest significant variation is between compressive strength (23.68 MPa) and permeability (10.18 mm/s) of PCP concrete with an aggregate size of 20–25 mm. This result demonstrates PCP concrete's low performance. It is unsuitable for use in the building of pedestrian walkways that require heavy load or sustain heavyweight. The ideal aggregate size for PCP concrete should result in the least amount of variance in compressive strength and permeability. PCP concrete with aggregate sizes ranging from 10 to 20 mm is the best aggregate size of PCP concrete since it has the least fluctuation in compressive strength and permeability.

Acknowledgements The authors would like to thank Swinburne University of Technology Sarawak Campus and Universiti Malaysia Sarawak (UNIMAS) for the collaboration.

References

Abd Elaty, M. A. A., & Ghazy, M. F. (2016). Evaluation of consistency properties of freshly mixed concrete by cone penetration test. *HBRC Journal, 12*(1), 1–12. https://doi.org/10.1016/j.hbrcj.2014.09.001

ACI Committee 522. (2010). *522R-10: Report on Pervious Concrete* (pp. 1–38). American Concrete Institute.

Ajamu, S. O., Jimoh, A. A., & Oluremi, J. R. (2012). Evaluation of structural performance of pervious concrete in construction. *International Journal of Engineering and Technology, 2*(5), 829–386.

Alimohammadi, V., Maghfouri, M., Nourmohammadi, D., Azarsa, P., Gupta, R., & Saberian, M. (2021). Stormwater runoff treatment using pervious concrete modified with various nano-materials: a comprehensive review. *Sustainability, 13*(15), 8552. https://doi.org/10.3390/su13158552

Alsayed, S. H., & Amjad, M. A. (1996). Strength, water absorption and porosity of concrete incorporating natural and crushed aggregate. *Journal of King Saud University—Engineering Sciences, 8*(1), 109–119. https://doi.org/10.1016/S1018-3639(18)30642-1

ASTM C1611/C1611M-21. (2021). *Standard test method for slump flow of self-consolidating concrete.* ASTM International. https://doi.org/10.1520/C1611_C1611M-21

ASTM C1701/C1701M-17a. (2017). *Standard test method for infiltration rate of in place pervious concrete*. ASTM International. https://doi.org/10.1520/C1701_C1701M-17A

ASTM C1723-16. (2016). *Standard guide for examination of hardened concrete using scanning electron microscopy*. ASTM International. https://doi.org/10.1520/C1723-16

ASTM C1747/C1747M-13. (2013). *Standard test method for determining potential resistance to degradation of pervious concrete by impact and abrasion*. ASTM International. https://doi.org/10.1520/C1747_C1747M-13

ASTM C944/C944M-19. (2019). *Standard test method for abrasion resistance of concrete or mortar surfaces by the rotating-cutter method*. ASTM International. https://doi.org/10.1520/C0944_C0944M-19

ASTM E1252-98. (2021). *Standard practice for general techniques for obtaining infrared spectra for qualitative analysis*. ASTM International. https://doi.org/10.1520/E1252-98R21

ASTM E1508-12. (2019). *Standard guide for quantitative analysis by energy-dispersive spectroscopy*. ASTM International. https://doi.org/10.1520/E1508-12AR19

ASTM E168-16. (2016). *Standard practices for general techniques of infrared quantitative analysis*. ASTM International. https://doi.org/10.1520/E0168-16

Barisic, I., Galic, M., & Grubesa, I. N. (2017). Pervious concrete mix optimization for sustainable pavement solution. *IOP Conference Series: Earth and Environmental Science, 90*(1), 012091. https://doi.org/10.1088/1755-1315/90/1/012091

Bechemeier, K. W. (2012). *Effects of high volumes of fly ash on cement paste*. Master Thesis. Missouri University of Science and Technology, pp. 1–235. https://scholarsmine.mst.edu/cgi/viewcontent.cgi?article=7941&context=masters_theses

Bonneau, J., Fletcher, T. D., Costelloe, J. F., & Burns, M. J. (2017). Stormwater infiltration and the 'urban karst'—A review. *Journal of Hydrology, 552*(1), 141–150. https://doi.org/10.1016/j.jhydrol.2017.06.043

BS EN 12390-3. (2019). *Testing hardened concrete. Compressive strength of test specimens*. British Standard Institution.

BS EN 197-1. (2014). *Cement, composition, specifications and conformity criteria for common cements*. British Standard Institution, London.

BS 882. (1992). *Specification for aggregates from natural sources for concrete*. British Standard Institution, London.

Chan, N., Young-Rojanschi, C., & Li, S. (2018). Effect of water-to-cement ratio and curing method on the strength, shrinkage and slump of the biosand filter concrete body. *Water Science & Technology, 77*(6), 1744–1750. https://doi.org/10.2166/wst.2018.063

Chin, M. Y., Rahman, M. R., Yong, C. W. P., Wong, A. M. C., & Bakri, M. K. B. (2020). The curing times effect on the strength of ground granulated blast furnace slag (GGBFS) mortar. *Construction and Building Materials, 260*(1), 120622. https://doi.org/10.1016/j.conbuildmat.2020.120622

Chin, M. Y., Rahman, M. R., Kuok, K. K., Chiew, W. Y., Bakri, M. K. B. (2020b). Characterization and impact of curing duration on the compressive strength of coconut shell coarse aggregate in concrete. *BioResources, 16*(3), 6057–6073. https://ojs.cnr.ncsu.edu/index.php/BioRes/article/view/BioRes_16_3_6057_Chin_Curing_Duration_Compressive_Strength_Coconut_Shell

Chui, X., Zhang, J., Huang, D., Gong, X., Liu, Z., Hou, F., Cui, S. (2016). Measurement of permeability and the correlation between permeability and strength of pervious concrete. *DEStech Transactions on Engineering and Technology Research, 1*(1), 885–892. https://doi.org/10.12783/dtetr/ictim2016/5580

Crouch, L. K., Pitt, J., & Hewitt, R. (2007). Aggregate effects on pervious portland cement concrete static modulus of elasticity. *Journal of Materials in Civil Engineering, 19*(7), 561–567. https://doi.org/10.1061/(ASCE)0899-1561(2007)19:7(561)

Dash, S., & Kar, B. (2018). Environment friendly pervious concrete for sustainable construction. *IOP Conference Series: Materials Science and Engineering, 410*, 012005. https://doi.org/10.1088/1757-899X/410/1/012005

Delatte, N., & Schwartz, S. S. (2010). Sustainability benefits of pervious concrete pavement. In *Proceedings of Honouree Sessions*, pp. 1–9.

Gardner, D., Lark, R., Jefferson, T., & Davies, R. (2018). A survey on problems encountered in current concrete construction and the potential benefits of self-healing cementitious materials. *Case Studies in Construction Materials, 8*(1), 238–247. https://doi.org/10.1016/j.cscm.2018.02.002

Goede, W., & Haselbach, L. (2011). Investigation into the structural performance of pervious concrete. *Journal of Transportation Engineering, 138*(1), 1–10. https://doi.org/10.1061/(ASCE)TE.1943-5436.0000305

Gupta, R. (2014). Monitoring in situ performance of pervious concrete in British Columbia—A pilot study. *Case Studies in Construction Materials, 1*(1), 1–9. https://doi.org/10.1016/j.cscm.2013.10.001

Hamdy, G. A. (2016). *Pervious concrete as a sustainable pavement solution—Overview of structural performance*. AICGSGE9, Egypt, pp. 1–8.

Hariyadi, T. H. (2015). Enhancing the performance of porous concrete by utilizing the pumice aggregate. *Procedia Engineering, 125*(1), 732–738. https://doi.org/10.1016/j.proeng.2015.11.116

Henrist, C., Mathier, J.-P., Vogels, C., Rulmont, A., & Cloots, R. (2003). Morphological study of magnesium hydroxide nanoparticles precipitated in dilute aqueous solution. *Journal of Crystal Growth, 249*(1–2), 321–330. https://doi.org/10.1016/S0022-0248(02)02068-7

Huang, J., Duan, T., Zhang, Y., Liu, J., Zhang, J., & Lei, Y. (2020). Predicting the permeability of pervious concrete based on the beetle antennae search algorithm and random forest model. *Advances in Civil Engineering, 1*, 1–11. https://doi.org/10.1155/2020/8863181

Hilpert, M., Mora, B. A., Ni, J., Rule, A. M., & Nachman, K. E. (2015). Hydrocarbon release during fuel storage and transfer at gas stations: environmental and health effects. *Current Environmental Health Reports, 2*(1), 412–422. https://doi.org/10.1007/s40572-015-0074-8

Ibrahim, A., Mahmoud, E., Yamin, M., & Patibandla, V. C. (2014). Experimental study on Portland cement pervious concrete mechanical and hydrological properties. *Construction and Building Materials, 50*(1), 525–529. https://doi.org/10.1016/j.conbuildmat.2013.09.022

Jang, J. G., Ahn, Y. B., Souri, H., & Lee, H. K. (2015). A novel eco-friendly porous concrete fabricated with coal ash and geopolymeric binder: heavy metal leaching characteristics and compressive strength. *Construction and Building Materials, 79*(1), 173–181. https://doi.org/10.1016/j.conbuildmat.2015.01.058

Jimma, B. E. (2014). Workability-integrated mixture proportioning method for pervious concrete. PhD Thesis. Clemson University, pp. 1–266. https://tigerprints.clemson.edu/cgi/viewcontent.cgi?article=2542&context=all_dissertations

Joshi, T., & Dave, U. (2021). Construction of pervious concrete pavement stretch, Ahmedabad, India—Case study. *Case Studies in Construction Materials., 1*(1), e00622. https://doi.org/10.1016/j.cscm.2021.e00622

Kia, A., Delens, J. M., Wong, H. S., & Cheeseman, C. R. (2021). Structural and hydrological design of permeable concrete pavements. *Case Studies in Construction Materials, 15*(1), e00564. https://doi.org/10.1016/j.cscm.2021.e00564

Kia, A., Wong, H. S., & Cheeseman, C. R. (2017). Clogging in permeable concrete: a review. *Journal of Environmental Management, 193*(1), 221–233. https://doi.org/10.1016/j.jenvman.2017.02.018

Kia, A., Wong, H. S., & Cheeseman, C. R. (2021). High-strength clogging resistant permeable pavement. *International Journal of Pavement Engineering, 22*(3), 271–282. https://doi.org/10.1080/10298436.2019.1600693

Kim, Y.-Y., Lee, K.-M., Bang, J.-W., & Kwon, S.-J. (2014). (2014) Effect of W/C Ratio on durability and porosity in cement mortar with constant cement amount. *Advances in Materials Sciences and Engineering, 1*, 1–11. https://doi.org/10.1155/2014/273460

Kovac, M., & Sicakova, A. (2017). Pervious concrete as a sustainable solution for pavements in Urban Areas. In *Environmental Engineering' 10th International Conference*, Vilnius Gediminas Technical University, Lithuania, pp. 1–7. https://doi.org/10.3846/enviro.2017.031

Kubba, S. (2010) Choosing materials and products. In: S. Kubba (ed.) *Green construction project management and cost oversight*. Architectural Press, Elsevier, pp. 221–266. https://doi.org/10.1016/B978-1-85617-676-7.00006-3

Kumar, L., Li, W. Z., Vannoy, C. H., Leblanc, R. M., & Wang, D. Z. (2009). Synthesis, characterization and optical properties of Mg(OH)2 micro-/nanostructure and its conversion to MgO. *Ceramic International, 32*(8), 3355–3364. https://doi.org/10.1016/j.ceramint.2009.05.035

Li, H., Harvey, J. T., Holland, T. J., & Kayhanian, M. (2013). The use of reflective and permeable pavements as a potential practice for heat island mitigation and stormwater management. *Environmental Research Letters, 8*(1), 049501. https://doi.org/10.1088/1748-9326/8/1/015023

Lee, M.-J., Lee, M.-G., Huang, Y., & Chiang, C.-L. (2012). Water purification of pervious concrete pavement. *International Conference on Sustainable Design, Engineering, and Construction.* https://doi.org/10.1061/9780784412688.089

Ma, K.-C., Chuang, M.-H., & Chan, T.-Y. (2021). Analysis of flooding adaptation and groundwater recharge after adopting J.W. ecological technology in a highly developed urbanization area. *Applied Sciences, 11*(6), 2662. https://doi.org/10.3390/app11062662

Matschei, T., Lothenbach, B., & Glasser, F. P. (2007). The role of calcium carbonate in cement hydration. *Cement and Concrete Research, 37*(4), 551–448. https://doi.org/10.1016/j.cemconres.2006.10.013

McGrane, S. J. (2016). Impacts of urbanization on hydrological and water quality dynamics, and urban water management: A review. *Hydrological Sciences Journal, 61*(13), 2295–2311. https://doi.org/10.1080/02626667.2015.1128084

Montes, F., Valavala, S., & Haselbach, L. (2005). A new test method for porosity measurements of Portland cement pervious concrete. *Journal of ASTM International, 2*(1), 1–13. https://doi.org/10.1520/JAI12931

Obla, K. H. (2010). Pervious concrete—an overview. *The Indian Concrete Journal, 84*(4), 9–18.

Offenberg, M. (2006). Proving pervious concrete's durability. Concrete Production & Forecast. Viewed at https://www.concreteconstruction.net/concrete-production-precast/proving-pervious-concretes-durability

Pilon, B. S., Tyner, J. S., Yoder, D. C., & Buchanan, J. R. (2018). The effect of pervious concrete on water quality parameters: a case study. *Water, 11*(1), 263. https://doi.org/10.3390/w11020263

Qin, Y., Yang, H., Deng, Z., & He, J. (2015). Water permeability of pervious concrete is dependent on the applied pressure and testing methods. *Advances in Materials Science and Engineering, 1*, 1–6. https://doi.org/10.1155/2015/404136

Rahman, M. A., Sarker, P. K., Shaikh, F. U. A., & Saha, A. K. (2017). Soundness and compressive strength of Portland cement blended with ground granulated ferronickel slag. *Construction and Building Materials, 140*(1), 194–202. https://doi.org/10.1016/j.conbuildmat.2017.02.023

Rodrigez-Hernandez, J., Andres-Valeri, V. C., Calzada-Perez, M. A., Vega-Zamanillo, A., & Castro-Fresno, D. (2015). Study of the raveling resistance of porous asphalt pavements used in sustainable drainage systems affected by hydrocarbon spills. *Sustainability, 7*(12), 16226–16236. https://doi.org/10.3390/su71215812

Sartipi, M., & Sartipi, F. (2019). Stormwater retention using pervious concrete pavement: great Western Sydney case study. *Case Studies in Construction Materials, 11*(1), e00274. https://doi.org/10.1016/j.cscm.2019.e00274

Sata, V., & Chindaprasirt, P. (2020). Use of construction and demolition waste (CDW) for alkali-activated or geopolymer concrete. In: Pacheco-Torgal, F., Ding, Y., Colangelo, F., Tuladhar, R., Koutamanis, A. (eds.) *Advances in construction and demolition waste recycling* (pp. 385–403). Woodhead Publishing. https://doi.org/10.1016/B978-0-12-819055-5.00019-X

Sata, V., Wongsa, A., & Chindprasirt, P. (2013). Properties of pervious geopolymer concrete using recycled aggregates. *Construction and Building Materials, 42*(1), 33–39. https://doi.org/10.1016/j.conbuildmat.2012.12.046

Schaefer, V. R., Wang, K., Suleiman, M. T., & Kevern, J. T. (2006). *Mix design development for pervious concrete in cold weather climates.* Ames: Iowa State University.

Shinde, G. U., & Valunjkar, S. S. (2015). Cost and time control in storm water management using pervious concrete. *International Journal for Scientific Research & Development, 3*(2), 1947–1949.

Shubbar, A. A., Jafer, H., Abdulredha, M., Al-Khafaji, Z. S., Nasr, M. S., Al Masoodi, Z., & Sadique, M. (2020). Properties of cement mortar incorporated high volume fraction of GGBFS and CKD from 1 day to 550 days. *Journal of Building Engineering, 30*(1), 101327. https://doi.org/10.1016/j.jobe.2020.101327

Spandana, K., & Rajasekhar, K. (2016). Strength properties of pervious concrete compared with conventional concrete. *IOSR Journal of Mechanical and Civil Engineering, 13*(4), 97–103. https://www.iosrjournals.org/iosr-jmce/papers/vol13-issue4/Version-3/P13040397103.pdf

Sonebi, M., Bassouni, M., & Yahia, A. (2016). Pervious concrete: Mix design, properties and applications. *RILEM Technical Letters, 2016*(1), 109–115. https://doi.org/10.21809/rilemtechlett.20

Tennis, P. D., Leming, M. L., & Akers, D. J. (2004). *Pervious concrete pavements, EB302.02*, Portland Cement Association, Skokie, Illinois, and National Ready Mixed Concrete Association, Silver Spring, Maryland, USA, pp. 1–36

Thives, L. O., Ghisi, E., Brecht, D. G., & Pires, D. M. (2018). Filtering capability of porous pavements. *Proceedings, 2*(5), 174. https://doi.org/10.3390/ecws-2-04943

Tho-in, T., Sata, V., Chindaprasirt, P., & Jaturapitakkul, C. (2012). Pervious high-calcium fly ash geopolymer concrete. *Construction and Building Materials, 30*(1), 366–371. https://doi.org/10.1016/j.conbuildmat.2011.12.028

Vaiciukyniene, D., Skipkiunas, G., Dauksys, M., & Sasnauskas, V. (2013). Cement hydration with Zeolite-based additive. *Chemija, 24*(4), 271–278.

Vinay, V., Adiseshu, S., & Chandramouli, S. Field application of pervious concrete for recharge of groundwater. *International Journal of Recent Technology and Engineering, 9*(2), 341–350. https://doi.org/10.35940/ijrte.B3203.07922

Voisin, J., Cournoyer, B., Vienney, A., & Mermillod-Blondin, F. (2018). Aquifer recharge with stormwater runoff in urban areas: Influence of vadose zone thickness on nutrient and bacterial transfers from the surface of infiltration basins to groundwater. *Science of the Total Environment, 637–638*(1), 1496–1507. https://doi.org/10.1016/j.scitotenv.2018.05.094

Vujovic, S., Haddad, B., Karaky, H., Sebaibi, N., & Boutouil, M. (2021). Urban heat Island: Causes, consequences, and mitigation measures with emphasis on reflective and permeable pavements. *CivilEng, 2*(2), 459–484. https://doi.org/10.3390/civileng2020026

Wanieliesta, M., & Chopra, M. (2007). *Performance assessment of Portland cement pervious pavement*. Final Report FDOT Project BD521-02. 1-80. https://www.fdot.gov/docs/default-source/roadway/drainage/Research/SA/FinalReport-1of4-Performance-Assessment-and-Credit.pdf

Wongsa, A. (2013). *The use of recycled materials in porous concrete*. Master Thesis. Khon Kaen University.

Wu, X.-F., Hu, G.-S., Wang, B.-B., & Yang, Y.-F. (2008). Synthesis and characterization of superfine magnesium hydroxide with monodispersity. *Journal of Crystal Growth, 310*(2), 457–461. https://doi.org/10.1016/j.jcrysgro.2007.10.025

Yaacob, I. I., Ali, M. Y., Sopyan, L., & Hashmi, S. (2015). Effect of calcium carbonate replacement on workability and mechanical strength of Portland cement concrete. *Advanced Materials Research, 1115*(1), 137–131. https://doi.org/10.4028/www.scientific.net/AMR.1115.137

Yao, A., Ding, H., Zhang, X., Hu, Z., Hao, R., & Yang, T. (2018). Optimum design and performance of porous concrete for heavy-load traffic pavement in cold and heavy rainfall region of N.E. China. *Advances in Materials Science and Engineering, 2018*(1), 1–15. https://doi.org/10.1155/2018/7082897

Yu, F., Sun, D., Wang, J., & Hu, M. (2019). Influence of aggregate size on compressive strength of pervious concrete. *Construction and Building Materials, 209*(1), 463–475. https://doi.org/10.1016/j.conbuildmat.2019.03.140

Zaetang, Y., Wongsa, A., Sata, V., & Chindaprasirt, P. (2015). Use of coal ash as geopolymer binder and coarse aggregate in pervious concrete. *Construction and Building Materials, 96*(1), 289–295. https://doi.org/10.1016/j.conbuildmat.2015.08.076

Zhang, R., Jiang, G., & Liang, J. (2015). The albedo of pervious cement concrete linearly decreases with porosity. *Advances in Materials Science and Engineering, 1*, 1–5. https://doi.org/10.1155/2015/746592

Zhu, H., Wen, C., Wang, Z., & Li, L. (2020). Study on the permeability of recycled aggregate pervious concrete with fibers. *Materials, 13*(2), 321. https://doi.org/10.3390/ma13020321

Glass Waste as Coarse Aggregate Filler Replacement in Concrete

Chin Mei Yun, Md Rezaur Rahman, Kuok King Kuok, Amelia Pei Sze Chai, Amanda Bong Shi Ding, and Muhammad Khusairy Bin Bakri

Abstract Glass has become necessary and irreplaceable in daily human life due to its benefits like abrasion resistance, durability, and availability in any form. The high demand for glass in our daily lives has resulted in a significant amount of glass waste. To address the issue of glass wastes, which has negative repercussions on individuals, culture, and the environment, glass waste is utilized as a partial replacement material for coarse aggregates in concrete manufacture. In earlier research, concrete made with glass waste of coarse particles improved concrete strength. Nonetheless, there have been few studies using glass waste as a superplasticizer in concrete. This study utilized glass waste as coarse particles in concrete, combined with a superplasticizer to boost its strength. An optimum glass waste amount as a coarse aggregate substitution is identified and optimized. Glass waste in the form of discarded glass bottles was broken into 5–20 mm fragments sizes and is used as coarse aggregates substitute in amounts of 5, 10, and 15%, respectively. The glass waste concrete samples were tested and evaluated for compressive strength after 7 and 28 days of curing to determine the concrete's strength. According to the research, 10% of the glass waste substitutes with conventional concrete result in equivalent pressure. Concrete's total compressive strength is 28.9 MPa when coarse particles are replaced with 10% glass waste. In addition, to enhance the strength of concrete, a superplasticizer was used with a 10% glass waste replacement. The concrete was treated with 0.8, 1, and 1.2 of superplasticizer, and it was discovered that the concrete treated with 0.8 ml had the highest compressive strength. Overall, coarse aggregates can be partially replaced with glass waste at a weight-to-weight ratio of up to 10%, but total replacement is not achievable. Still, more research into the influence of glass waste as coarse aggregates in concrete mixing on other characteristics is needed.

Keywords Concrete · Glass · Waste · Compressive · Strength

C. M. Yun · K. K. Kuok · A. P. S. Chai · A. B. S. Ding
Faculty of Engineering, Computing and Science, Swinburne University of Technology Sarawak Campus, Jalan Simpang Tiga, 93400 Kuching, Sarawak, Malaysia

M. R. Rahman (✉) · M. K. B. Bakri
Faculty of Engineering, Universiti Malaysia Sarawak, Jalan Datuk Mohammad Musa, 94300 Kota Samarahan, Sarawak, Malaysia
e-mail: rmrezaur@unimas.my

M. R. Rahman et al. (eds.), *Waste Materials in Advanced Sustainable Concrete*, Engineering Materials, https://doi.org/10.1007/978-3-030-98812-8_2

1 Introduction

Concrete is one of the most often utilized building materials nowadays. Natural resources have been depleted because of the global production of roughly 7.3 billion tons of concrete (Kanojia & Jain, 2017). The proper disposal of stored trash helps to safeguard the environment while also lowering construction expenses. According to Meldrum-Hannada et al. (2017) from ABC News Australia, consumers consume around 1.36 million tons of Glass each year. Despite the increased usage of Glass, the percentage of glass recycling has decreased from 49 to 42% (Industry Edge Pty Ltd. & Equilibrium OMG Pty Ltd., 2017).

The single-use nature of Glass for applications such as beverage bottles, one of its primary uses, is a significant issue. In 2016, discarded glass was expected to account for 5% of worldwide municipal solid trash (Kaza & Yao, 2018), with recycling rates varied internationally across regions. Glass recycling rates in Europe averaged 71.48% in 2017, with individual nations ranging from 98% (Slovenia and Belgium) to 9% (Turkey) (FEVE Recycling Statistic, 2017). In 2017, the United States had a glass recycling rate of 26.63 percent, with 52.9% of glass containers going to landfill. Glass waste is non-biodegradable, taking up valuable landfill space indefinitely. Due to poor recycling methods, the dumping of glass trash into landfills increases reliance on natural resources, depleting sources such as beaches to create additional glass goods (Harrison et al., 2020). As the demand for landfill space grows, landfill taxes are expected to encourage more significant recycling behaviors. Finding new ways to reuse discarded Glass may save money on disposal while also preserving landfills and natural resources (Harrison et al., 2020).

Glass may be reused in two ways: closed-loop or open-loop, as an alternative to landfill. Glass may be recycled in a closed-looped system by either recycling the glass containers through a return program or producing new glass goods from recycled glass cullet (Dyer, 2014). Carbon emissions and energy usage decreased by 5 and 3%, respectively, for every 10% of glass cullet utilized (by weight), especially in the glass manufacturing industries (Harrison et al., 2020). Glass cullet is the best type of glass recycling since the glass may be recycled indefinitely if contamination, such as food waste or cross-color contamination, is avoided (Geueke et al., 2018). The glass product and its quality are assured as the re-processors define permissible contamination limits in the glass cullet. Because additives in amber and green container glass can degrade the quality of clear container glass or discolor the final product, it is critical to avoid combining different colored glasses (Harrison et al., 2020). As a result, color sorting glass containers, which distinguish between clear, green, and amber glass containers, is crucial in the glass recycling process. In an open-looped system, glass waste is recycled as alternate goods before being discarded at the end of its useful life (Dyer, 2014). Ceiling insulation, usually constructed of glass fibers, is one example of such materials (Christensen & Damgaard, 2010). Open-loop systems are the least optimal kind of recycling since they stop the recycling process from continuing. Other types of glass recycling, on the other hand, are frequently unable to progress if the glass trash is not color-separated or if the quality is damaged.

Another application of open-loop recycling for waste glass is in the building sector, where cement-based products can be used. The waste glass was utilized as a partial clinker replacement in cement production, finely ground glass for pozzolanic characteristics in Portland-cement blends, or a filler replacement in cement-based products as fine or coarse aggregate in mortar or mortar concrete (Harrison et al., 2020). The cement sector is determined to be responsible for 8% of world emissions, with clinker production accounting for 90% of it (Seifan et al., 2016, 2020). Portland cement is made by mixing powdered clinker with gypsum, which regulates the cement's setting pace (Biernacki et al., 2017; Coppola et al., 2018; Dunuweera & Rajapakse, 2018). The process of making clinker produces a lot of pollution since it takes a lot of heat to assist calcination, which accounts for half of the emissions in cement manufacturing (Fayomi et al., 2019; Naqi & Jang, 2019).

For raw meal or paste production, proper chemical proportions of primary limestone ($CaCO_3$) and clay (SiO_2 and Al_2O_3) are processed and combined with additional minerals like shale (Fe_2O_3) (Sprung, 2008). The raw meal or paste is then roasted at high temperatures (900 °C) in a kiln to allow the calcination of the $CaCO_3$ to form calcium oxide (CaO) while releasing carbon dioxide (CO_2) as a by-product (Mohamed et al., 2012). Temperatures are raised to 1450 °C to allow calcium silicates ($3CaO.SiO_2$, C_3S; and $2CaO.SiO_2$, C_2S), tricalcium aluminate ($3CaO.Al_2O_3$, C_3A), and tetra-calcium alumino-ferrite ($4CaO.Al_2O_3.Fe_2O_3$, C_4AF) to form and partly molten to produce clinker grain.

The waste glass was used in cement-based products to create a more sustainable future (Abdelli et al., 2020; Al-jbudri et al. 2019; Harrison et al., 2020; Ogundairo et al., 2019; Qin et al., 2021). Currently, there are few commercial experiments of waste glass being utilized in cement-based products. The Olympic Delivery Authority (ODA) has prioritized sustainability to reduce the concrete carbon footprint by 25% (Henson, 2011). Glass sand replaces the fine aggregate replacement rates ranging from 0 to 15%, depending on the required strength designation. In contrast, the fine aggregate was replaced with glass sand at a rate of 2% on average. (Harrison et al., 2020). The high silica component of glass, on the other hand, created the possibility of an alkali-silica reaction (ASR), which can cause concrete to expand and break. To avoid this, it took a year of lab study and testing to figure out the proper mix proportions for the project to be completed successfully.

Because glass is perishable and harmful to the environment, the remainder is being stored in the landfill. As the production and usage of concrete grow, replacing glass waste with coarse aggregates can assist in solving the issue of natural resource degradation and scarcity. In concrete, water reduction also resulted in good density, pasted consistency, stronger compressive and flexural strength because the water-to-cement ratio significantly influences the characteristics of concrete. As a result, a water-reducing additive such as a superplasticizer was added to the concrete mix. The ideal percentage of the glass waste was mixed with the naphthalene sulfonate formaldehyde superplasticizer. A plasticizer has been utilized and incorporated with a coarse aggregate substitute in almost every concrete. Even though the studies on the use of glass waste as coarse aggregates in concrete have been conducted in recent years, this study looked at the impacts of adding superplasticizer to glass trash as coarse particles in concrete. Experimental testing and characterization were used to determine the compressive strength of concrete and identify its influence on the strength. The use of glass trash as coarse aggregate was thought to increase the compressive strength of concrete. The most optimal and best proportion of glass wastes to be utilized as coarse aggregate replacements in concrete has been calculated to be around 10%.

2 Concrete with Glass Waste as Coarse Aggregates

A few studies on the workability of concrete with glass waste as partial substitution of coarse aggregates have been conducted. Replacement of coarse aggregates with glass waste does not affect the slump and workability, according to Shayan and Xu (2004). However, Ismail and Al-Hashimi (2009) found that the possibility for slump decreases as the percentage of glass waste increases. Therefore, this is primarily due to poor glass geometry, resulting in reduced workability and fineness modulus (Nafisa, 2020; Park et al., 2004).

A few studies show that glass as coarse aggregate in concrete workability decreases as the amount of glass substantial increases in the concrete (Rachael et al., 2019). In addition, it is also due to discarded glass's surface roughness and its inability to retain water (Rachael et al., 2019). Singh et al. (2015) concurred with the preceding reasoning that the decrease in the slump is caused by the loss of successful water from the specimen during rapid transit, caused by the waste glass component creating more voids. Furthermore, according to Singh et al. (2015), the waste glass may be utilized to replace coarse aggregates, with a replacement ratio of 10% waste glass as coarse aggregates being suitable. It was also reported that a more significant amount of replacement leads to poorer concrete compressive strength. Furthermore, few researchers discovered that increasing waste glass lowers sample compressive strength during curing (Rachael et al., 2019). The more significant the amount of waste glass in a concrete mix, the lower the concrete's compressive strength.

Waste glass may be utilized as coarse aggregate in concrete, improving the looks but lowering the mechanical properties. The compressive and splitting tensile strengths were significantly reduced when natural coarse aggregate was replaced with coarse waste glass, according to Afshinnia and Rangaraju (2016), with a loss of 38% of the compressive strength. Few researchers demonstrated that substituting the glass waste for typical coarse aggregate reduced the compressive and splitting tensile strengths while having minimal impact on workability (Kou & Poon, 2009; Topcu & Canbaz, 2004). Because of the Glass flat surface, a weaker connection occurs between the crushed glass aggregate and the cement paste, reducing the produced strength. According to Gerges et al. (2018), increasing the coarse waste glass replacement ratio lowers the 7th-day and 28th-day compressive strengths. Compared to the control sample, the samples containing coarse waste glass had a more significant differential in compressive strength at 7-days than at 28-days. In addition, this shows that the waste glass samples exhibit late strength development. Terro (2006) discovered that replacing coarse aggregate by up to 10 vol% can result in higher compressive strength at temperatures over 150 °C, but replacing coarse aggregate by more than 10% produces a loss in compressive strength at extreme temperatures. The author also discovered that using scrap glass as a fine aggregate yields better results than coarse aggregate. According to Topcu and Canbaz (2004), adding waste glass aggregate to the mix enhanced the Ve-Be consistency test results. The inclusion of waste glass has minimal impacts on the workability, resulting in a decrease in a slump but an improvement in uniformity. It is discovered that an increase in droop due to the smoothness of the glass surface results in less interference with the mix (Terro, 2006).

3 Concrete with Glass Waste as Coarse Aggregates and Addition of Superplasticizer

According to Matias et al. (2014), applying superplasticizers to reclaim concrete increases its performance. Adding a superplasticizer to concrete reduces the pattern of concrete workability. It was determined that superplasticizer might enhance and strengthen the compressive strength of concrete, but that increasing the dose decreases the compressive strength (Matias et al., 2014). However, a small number of scholars disagree with this claim (Faroug et al., 1999). Superplasticizers are said to improve workability by decreasing shear and flow resistance in concrete. Furthermore, adding a superplasticizer to concrete lengthens the time it takes to lose its workability (Faroug et al., 1999).

According to Mohammed et al. (2016), increasing the superplasticizer dosage improves the compressive strength of concrete. However, there is an ideal dose for adding superplasticizer since increasing the dosage beyond the limit would reduce the compressive strength of concrete. Overdosing with superplasticizer resulted in bleeding and segregation, affecting the cohesion and homogeneity of concrete (Mohammed et al., 2016).

4 Methodology

4.1 Materials

"Heineken" brand bottle waste was obtained from local restaurants. Portland limestone cement and sand were obtained from Cahya Mata Sarawak (CMS) Cement Sdn. Bhd. Kao Industrial (Thailand) Co. supplied the naphthalene sulfonate formaldehyde superplasticizer. Ltd and contains naphthalene sulfonic acid, sodium salt, water, and a formaldehyde polymer. River sand was sourced from a local supplier in Sarawak and used as fine aggregates.

4.2 Materials Preparation

As glass waste, a total of 60 "Heineken" bottles were collected and smashed into tiny pieces. The crushed glass pieces were then utilized to conduct sieve analysis. By substituting the coarse aggregates in the concrete, the coarse waste glass had a particle size diameter ranging from 4.75 to 20 mm. This study employed Portland Limestone Cement (PLC) as a binder in the glass waste coarse aggregate concrete manufacturing process.

Coarse aggregates are made up of rocks of varying sizes. The gravels were sieved to separate the sizes since they were utilized as coarse aggregates. Figure 1 shows the results of sieve analysis for coarse aggregates. During the investigation, fine aggregates were made from river sand with a maximum size of 4.75 mm. The sand was sieved to remove larger sand particles. To identify the optimum proportion of glass wastes to utilize as coarse aggregate substitutes in concrete, the first part of the studies investigated waste glass substitution of coarse aggregates at 5, 10, and 15%, respectively. Other studies used different amounts of superplasticizer, which is naphthalene sulfonate formaldehyde. After establishing the optimal proportion of glass waste as coarse aggregate replacements in concrete, they were added to the concrete samples. With 10% glass waste substitution, superplasticizers were added to the concrete at 0.8, 1.0, and 1.2%, respectively.

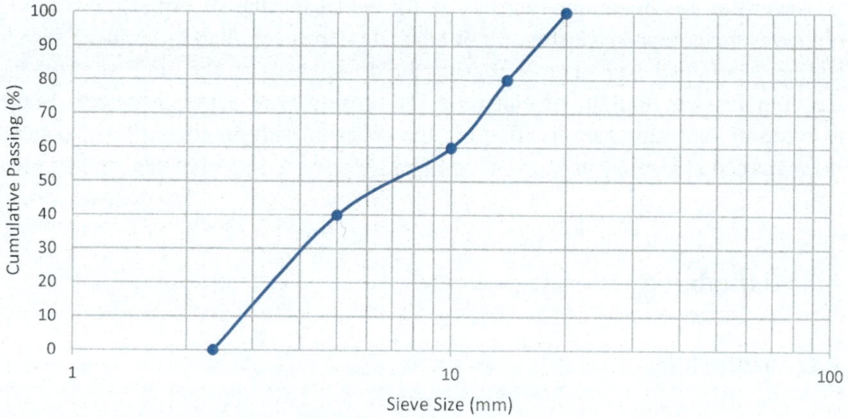

Fig. 1 Coarse aggregate sieve analysis

Table 1 Design of concrete mix for the control sample

Material	Cement	Water	Coarse aggregates	Fine aggregates
Amount of material required (kg/m^3)	400	250	1100	750

Table 2 Concrete mix design of all samples (Phase 1)

Specimen	Cement (kg/m^3)	Water (kg/m^3)	Coarse aggregates (kg/m^3)	Fine aggregates (kg/m^3)	Glass waste replacement (%)	Glass waste replacement (kg/m^3)
OPC	400	200	1100	730	0	0
WG5	400	200	1045	730	5	55
WG10	400	200	990	730	10	110
WG15	400	200	935	730	15	165

4.3 Concrete Mix Design

The correct amounts of cement, water, and aggregates were utilized to produce workable concrete with the desired and required strength. For the first part of the investigation, four concrete mixes with 0, 5, 10, and 15% glass waste replacement as coarse aggregates were cast and tested. Table 1 shows the concrete mix design for the control sample, whereas Table 2 shows the concrete mix design for all analyses. In addition to the first part of the investigation, for the second part, an additional concrete batch was made with 10% glass waste as a coarse aggregate alternative, and a superplasticizer was added. With a 10% glass waste substitution, the concrete receives 0.8, 1.0, and 1.2 ml of superplasticizers, respectively. Table 3 shows the concrete mix configuration with the addition of a superplasticizer.

4.4 Concrete Casting, Curing, Testing, and Characterization

The materials were mixed once they had been prepared and weighed. After each batch of concrete was mixed, a workability test was carried out to determine its workability through a slump test on the concrete mix. Each batch of concrete mix design consists of six 100 mm × 100 mm × 100 mm concrete cubes. After 24 h, the concrete cubes were de-molded and submerged in water to cure. Compressive strength tests were repeated following a 7-day and 28-day curing period to assess the concrete's compressive strength.

Table 3 Concrete mix design with the addition of superplasticizer (Phase 2)

Specimen	Cement (kg/m^3)	Water (kg/m^3)	Coarse aggregates (kg/m^3)	Fine aggregates (kg/m^3)	Glass waste replacement (%)	Glass waste replacement (kg/m^3)	Superplasticizer (% of total cementitious weight)
WG10SP0.8	400	180	990	730	10	0.88	0.8
WG10SP1.0	400	180	990	730	10	0.88	1
WG10SP1.2	400	180	990	730	10	0.88	1.2

4.4.1 Slump Test

Before casting, the slump test was performed after the mixing procedure to ensure that each concrete batch was workable. ASTM C142/C142M-20 (2020) standard was used as a reference.

4.4.2 Compressive Test

The compressive strength test was conducted under BS EN 12,390-3 (2019) using Unit Test Autocon 2000. The load was gradually applied to the mortar specimens at 0.05 ± 0.01 kN/s over an area of 100 mm × 100 mm during the compressive strength test until the samples failed. Results obtained from mortar specimens of the same batch were averaged and recorded.

4.4.3 Scanning Electron Microscope (SEM) and Energy Dispersive Spectroscopy (EDS/EDX)

Scanning electron microscopy (SEM) and energy dispersive spectroscopy (EDX/EDS) were carried out in compliance with ASTM C1723-16 (2016) and ASTM E1508-12 (2019) standards, respectively. 100×, 250×, 500×, and 10,000× magnifications were used to analyze the samples. Automated software scans and analyses the elemental composition percentages of the samples. The EDX/EDS was conducted on each sample many times at various phases, with the most representative results chosen. Hitachi TM4000Plus Tabletop Microscope (Hitachi Ltd., Tokyo, Japan) with a Quantax75TM Series Energy Dispersive X-Ray Spectrometer was used for the SEM and EDX/EDS investigations of the materials.

4.4.4 Fourier Transform Infrared (FTIR) Spectroscopy

A Fourier-transform infrared (FTIR) spectroscopy examined the materials (IRAffinity-1, Shimadzu Corporation, Kyoto, Japan). Fourier-transform infrared spectroscopy was used for qualitative and quantitative analysis under ASTM E168-16 (2016) and ASTM E1252-98 (2021) standards. The spectra of each sample were scanned in the range of 4000–400 cm^{-1}. Fourier-transform infrared spectroscopy is used for the samples, whereas the infrared spectrum transmittance and absorption were utilized to generate a unique chemical fingerprint spectrum. The test was repeated multiple times for each sample, and the most representative results were chosen.

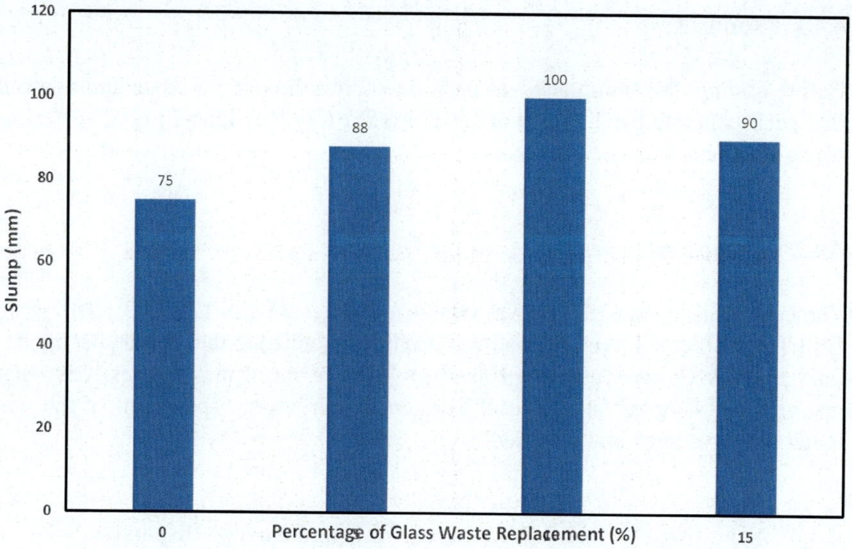

Fig. 2 Slump for glass waste replacement with different percentages (Phase 1)

5 Results and Discussions

5.1 Workability/Slump Properties

5.1.1 Concrete with Various Percentages of Glass Waste Replacement as Coarse Aggregates

Figure 2 depicts the results of the concrete slump test achieved during the laboratory test. A concrete mix's optimum slump value should be between 75 and 100 mm. Based on the results, the slump test of concrete using glass waste as coarse aggregates substitution is 75, 88, 90, and 100 mm, respectively. Both concrete samples meet the building mix requirements and are thus considered workable. Similar results were shown by many researchers (Hoang & Pham, 2016). Suppose the optimum workability for concrete is not achieved. In that case, the properties of consistency, flowability, compatibility, and harshness of concrete might be affected, leading to a possible crack or defects (Hoang & Pham, 2016; Li et al., 2016; Oztas et al., 2006).

5.1.2 Concrete with 10% Glass Waste Replacement as Coarse Aggregates and Addition of Superplasticizer

A slump test was conducted after the ingredients were mixed to assess the concrete's workability, and the results were summarized in Fig. 3. The value of the slump test

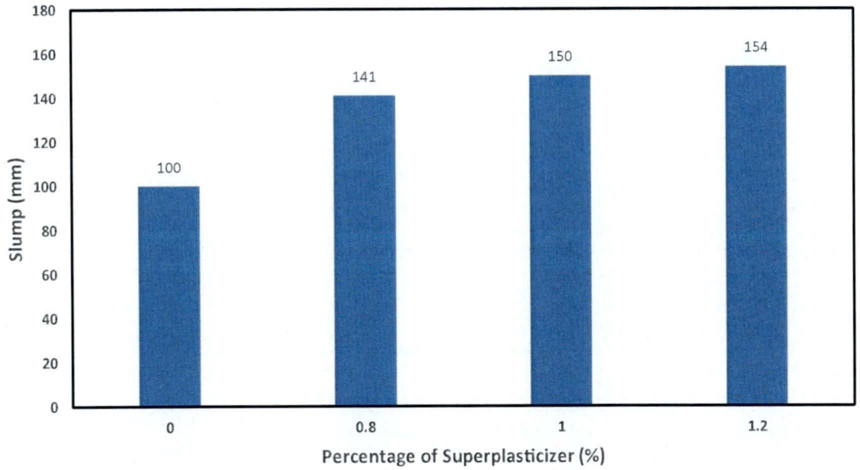

Fig. 3 Slump for superplasticizer with different percentages (Phase 2)

improves with the use of a superplasticizer, according to the data. Superplasticizer works as a water reducer, lowering the quantity of water required for the concrete mix as the amount of superplasticizer is increased. As the amount of superplasticizer is raised, it acts as a water reducer, reducing the amount of water available for the concrete mix. The correct amount of superplasticizer resulted in a loss of compressive strength, despite enhancing its workability (Malhotra, 1981; Salem et al., 2016).

5.2 *Compressive Strength Properties*

5.2.1 Concrete with Various Percentages of Glass Waste Replacement as Coarse Aggregates

The results of the concrete compressive strength test after 7 and 28 days are shown in Fig. 4. The average compressive strength of three concrete cubes was computed during the assessment to get the findings. Figure 4 shows that the compressive strength of concrete improves as the proportion of glass waste substituted grows until the optimum ratio is reached, beyond which it begins to fall. Concrete with 10% glass waste as coarse particles has greater compressive strength than ordinary concrete. According to Singh et al. (2015), the increasing amount of glass wastes used as coarse particles in concrete decreases the hardness of the concrete. As the percentage of glass waste is raised, concrete shrinkage is minimized. The inability of the glass to hold water prevents energy from being released during the hydration reaction, resulting in this result. Another factor that might influence the drop in the compressive strength was the surface morphology of the coarse aggregate (Tittarelli et al., 2018). The

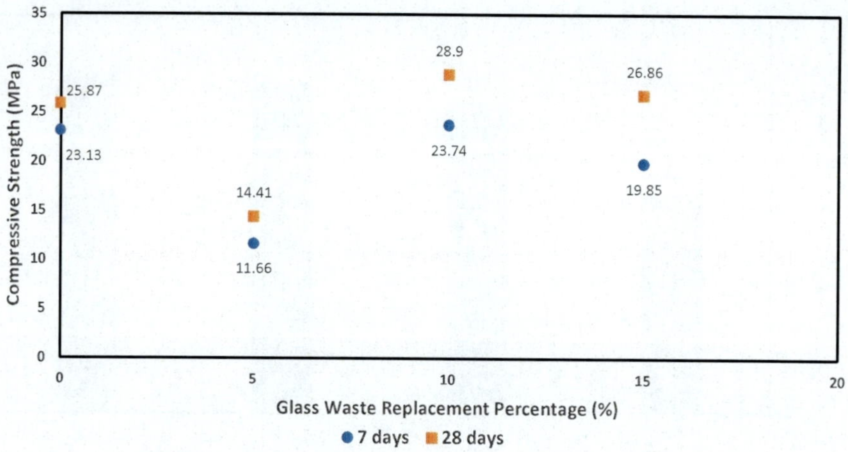

Fig. 4 Compressive strength with different glass waste replacement (Phase 1)

presence of glass waste in the cementitious composite may provoke the possibility of its expansion and crack. The amorphous silica in glass dissolved under alkaline conditions and form alkali-silica reaction (ASR) gel, which absorbs water and subsequently expands, leading to irregular cracks, which reduce the durability of concrete (Bignozzi et al., 2015; Rashad, 2014, 2015).

5.2.2 Concrete with 10% Glass Waste Replacement as Coarse Aggregates and Addition of Superplasticizer

Figure 5 shows the results of a concrete compressive strength test after 7 and 28 days of superplasticizer addition. Compared to the normal concrete, the compressive strength of the concrete with the added percentage of superplasticizers improved alongside the addition of plasticizers, as shown in Fig. 5. Increasing the amount of superplasticizer, on the other hand, lowers the compressive strength of concrete. According to Mohammed et al. (2016), concrete containing a coarse aggregate replacement of 10% glass waste and a superplasticizer of 0.8% obtains the highest compressive strength. Overdosing with superplasticizer produces bleeding and segregation, reducing the cohesion and homogeneity of the concrete. Another reason for it was an improper reaction between the plasticizer and concrete substance. In contrast, it would create an on-spot globular response, which is not appropriately dispersed in the concrete (Rashad, 2014, 2015).

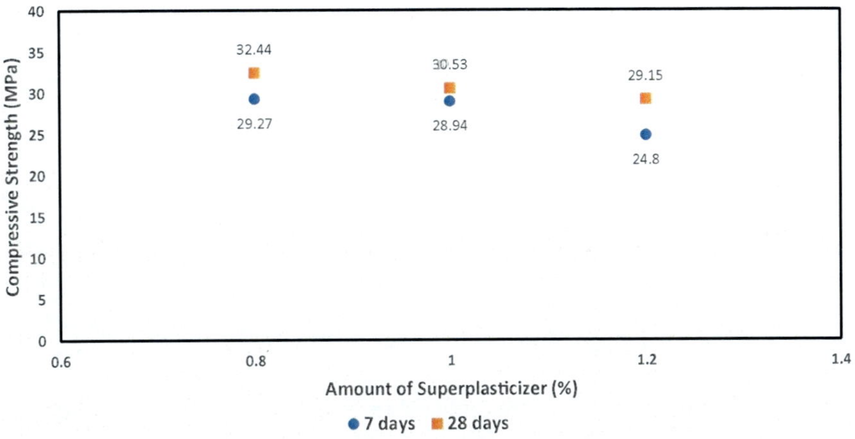

Fig. 5 Compressive strength with different percentages of superplasticizer (Phase 2)

5.3 Morphological Properties

SEM pictures of the WG10 (7 days) are shown in Fig. 6a–d at magnifications of 100x, 250x, 500x, and 1000x, respectively. The topology of the surface sample was flakily mixed with unreacted structure, which is very weak, as shown in Fig. 6. The hydraulic reactivity, or chemical or physical interaction, was low to generate strong interlock, preventing it from being combined within themselves or with other materials (Al-jburi Najad et al., 2019). According to other researchers, to create proper interlock, the appropriate paste combination, soundness, and hydraulic reactivity were responsible for the compact microstructure (Chin et al. 2020a, 2020b; Henrist et al. 2003; Kumar et al. 2009; Shubbar et al. 2020; Wu et al. 2008). Furthermore, Figs. 6c, d show small traces of reaction between the amorphous silica in glass, partly dissolved under alkaline conditions to form alkali-silica reactions. Few researchers have supported these findings (Bignozzi et al., 2015; Rashad, 2014, 2015).

5.4 Elemental Analysis

Tables 4 and Fig. 7 show the EDS/EDX plot and element contents for WG10 (7 days). The chemical elements present in the sample, according to Table 4, were carbon (C), oxygen (O), and calcium (Ca), silicon (Si), aluminum (Al), iron (Fe), and sulfur (S). Table 4 reveals that the most significant content is the mass oxygen percentage, followed by calcium and carbon. It also demonstrates that a chemical reaction occurs during the mixing paste and hydraulic reactivity, resulting in calcium carbonate,

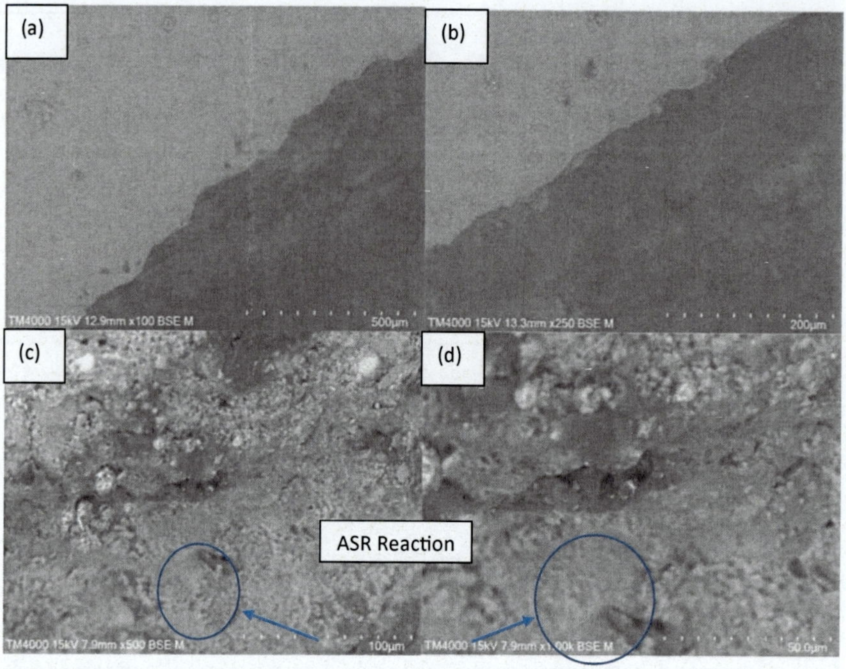

Fig. 6 SEM images for WG10 (7 days) at magnification of **a** 100×, **b** 250×, **c** 500×, and **d** 1000×

Table 4 Element composition for WG10 (7 days)

Element	Mass normal (%)	Atom (%)
O	52.06	57.54
Ca	34.27	15.12
C	13.77	20.27
Si	8.13	5.12
Al	1.49	0.98
Fe	1.40	0.44
S	0.98	0.54
Sum	100.00	100.00

$CaCO_3$ (Matschei et al., 2007; Yaacob et al., 2015). $CaCO_3$ helps to improve early strength by having an accelerator effect and a high rate of hydration, which causes the concrete to harden faster (Matschei et al., 2007; Yaacob et al., 2015). Another chemical reaction revealed the presence of glass in the concrete as SiO_2, Al_2O_3, and Fe_2O_3 (Islam et al., 2017).

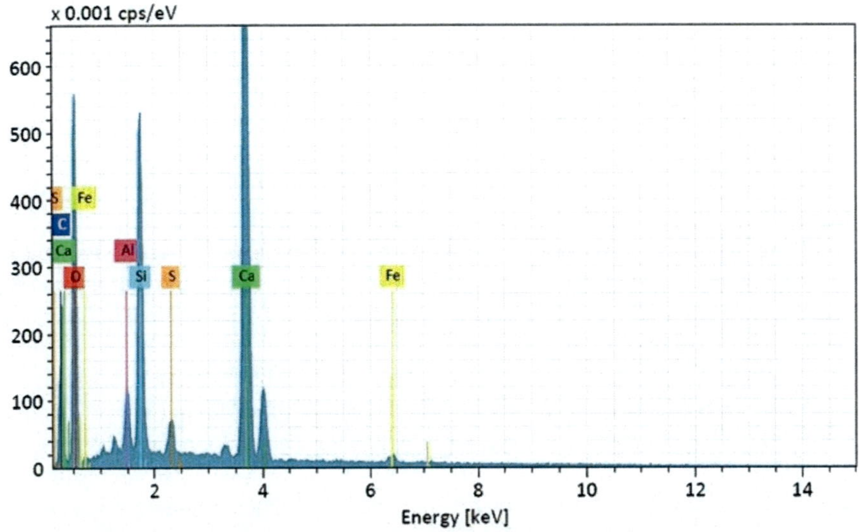

Fig. 7 EDS/EDX graph for WG10 (7 days)

5.5 *Infrared Spectral Properties*

Figure 8 shows the FTIR images for WG10 (7 days). The small broad peak at the 3200-3500 cm^{-1} band was due to the O–H bond due to the hydraulic of water bond and moisture. The peak band for 1670.35, 1390.68, 627.46, 677.01, and 582.50 cm^{-1} was observed to be Si–O–Si and stretched Si–OH. It is noted that polymerization of Si–OH to Si–O–Si also takes place in the setting of the cement (Thoo et al., 2013).

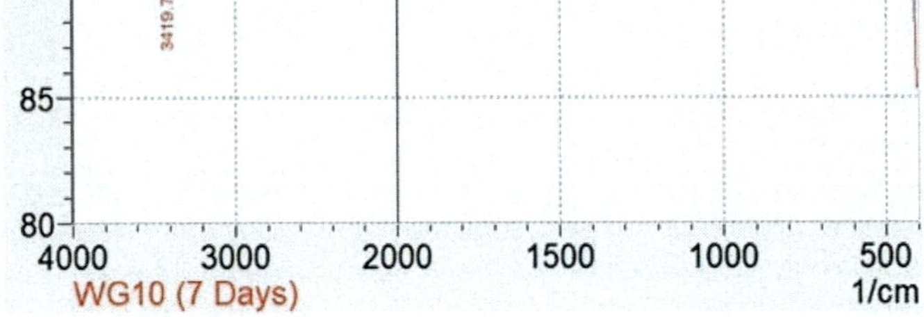

Fig. 8 FTIR for WG10 (7 days)

6 Utilizing Coarse Glass Waste in Concrete

In making concrete, Cheeseman (2011) used leftover glass to substitute the coarse glass material partially and discovered a decrease in workability and compressive strength. Other studies found no substantial effect of coarse glass aggregate on concrete workability, only a slight drop in strength (Topcu & Canbaz, 2004). Glass cullet was utilized as coarse aggregate in concrete by Topçu et al. (2008) and Palmquist (2003). Because of alkali-silica reaction (ASR), they discovered that glass aggregate was producing expansions in concrete or mortar and internal strains, resulting in a loss of durability, cracked formation, and increased permeability. Glass cullet was also utilized as aggregate in concrete.

Abdullah (2011) discovered that the concrete unit weight was reduced when coarse aggregate glass exceeded 40%. In addition, the coarse glass did not affect the mix's workability. The numbers were run repeatedly, and the best content percentage of glass coarse aggregate to use as a partial substitute for natural coarse aggregate with a w/c ratio of 0.4 or 40% was about 0.265, with a 28-day compressive strength of around 385 kg/cm^2. It also found no influence on pull-out strength, improvement in flexural strength, and a minor drop in the splitting strength of concrete.

7 Conclusion

According to the Environmental Council of the Concrete Organization, this concrete made from glass waste can be utilized in support layers and sub-bases (EECO). Glass trash is used to partially cover coarse particles in concrete, according to the study. The idea that coarse aggregates made up of 10% glass waste give the maximum compressive intensity, supported by other characterization. 0.8% of the superplasticizer is added to the concrete to increase the workability and compressive strength. The usage and use of recovered glass trash as coarse aggregates can also help conserve natural resources by reducing discarded glass by creating fewer environmental problems.

Acknowledgements The authors would like to thank Swinburne University of Technology Sarawak Campus and Universiti Malaysia Sarawak (UNIMAS) for the collaboration.

References

Abdelli, H. E., Mokrani, L., Kennouche, S., & Barroso de Aguiar, J. L. (2020). Utilization of waste glass in the improvement of concrete performance: A mini review. *Waste Management & Research: the Journal for a Sustainable Circular Economy, 38*(11), 1204–1213. https://doi.org/10.1177/0734242X20941090

Abdullah, A. S. (2011). *Properties of concrete mixes with waste glass.* Master Thesis. Civil Engineering Rehabilitation and Design of Structures.

Afshinnia, K., & Rongaraju, P. R. (2016). Impact of combined use of ground glass powder and crushed glass aggregate on selected properties of Portland cement concrete. *Construction and Building Materials, 117*(1), 263–272. https://doi.org/10.1016/j.conbuildmat.2016.04.072

Al-jburi Najad, A. A., Hasan Kareem, J. K., Azline, N. l., & Ostovar, N. (2019). Waste glass as partial replacement in cement—A review. *IOP Conference Series: Earth and Environmental Science, 357*(1), 012023. https://doi.org/10.1088/1755-1315/357/1/012023

ASTM C143/C143M-20. (2020). *Standard test method for slump of hydraulic-cement concrete.* ASTM International. https://doi.org/10.1520/C0143_C0143M-20

ASTM C1723-16. (2016). *Standard guide for examination of hardened concrete using scanning electron microscopy.* ASTM International. https://doi.org/10.1520/C1723-16

ASTM E1252-98. (2021). *Standard practice for general techniques for obtaining infrared spectra for qualitative analysis.* ASTM International. https://doi.org/10.1520/E1252-98R21

ASTM E1508-12. (2019). *Standard guide for quantitative analysis by energy-dispersive spectroscopy.* ASTM International. https://doi.org/10.1520/E1508-12AR19

ASTM E168-16. (2016). *Standard practices for general techniques of infrared quantitative analysis.* ASTM International. https://doi.org/10.1520/E0168-16

ASTM E1252-98. (2013). *Standard practice for general techniques for obtaining infrared spectra for qualitative analysis.* ASTM International. https://doi.org/10.1520/E1252-98R13E01

Biernacki, J. J., Bullard, J. W., Sant, G., Brown, K., Glasser, F. P., Jones, S., Ley, T., Livingston, R., Nicoleuau, L., Olek, K., Sanchez, F., Shahsavari, R., Stutzman, P. E., Sobolev, K., & Prater, T. (2017). Cements in the 21st century: Challenges, perspectives, and opportunities. *Journal of the American Ceramic Society, 100*(7), 2746–2773. https://doi.org/10.1111/jace.14948

Bignozzi, M. C., Saccani, A., Barbieri, L., & Lancellotti, I. (2015). Glass waste as supplementary cementing materials: The effects of glass chemical composition. *Cement and Concrete Composites, 55*(1), 45–52. https://doi.org/10.1016/j.cemconcomp.2014.07.020

BS EN 12390-3. (2019). *Testing hardened concrete. Compressive strength of test specimens.* British Standard Institution

Cheeseman, C. (2011) Production of sintered lightweight aggregate using waste ash and other industrial residues. In *Proceedings of the Second International Slag Valorisation Symposium: The Transition to Sustainable Materials Management* (pp. 89–101). Katholieke Universiteit Leuven.

Chin, M. Y., Rahman, M. R., Yong, C. W. P., Wong, A. M. C., & Bakri, M. K. B. (2020a). The curing times effect on the strength of ground granulated blast furnace slag (GGBFS) mortar. *Construction and Building Materials, 260*(1), 120622. https://doi.org/10.1016/j.conbuildmat.2020.120622

Chin, M. Y., Rahman, M. R., Kuok, K. K., Chiew, W. Y., & Bakri, M. K. B. (2020b). Characterization and impact of curing duration on the compressive strength of coconut shell coarse aggregate in concrete. *BioResources, 16*(3), 6057–6073. https://ojs.cnr.ncsu.edu/index.php/BioRes/article/view/BioRes_16_3_6057_Chin_Curing_Duration_Compressive_Strength_Coconut_Shell

Christensen, T. H., & Damgaard, A. (2010). Recycling of glass. In T. H. Christensen (Ed.) *Solid waste technology and management.* Wiley and Sons. https://doi.org/10.1002/9780470666883.ch16

Coppola, L., Bellezze, T., Belli, A., Bignozzi, M. C., Bolzoni, F., Brenna, A., Cabrini, M., Candamano, S., Cappai, M., Caputo, D., Carsana, M., Casndi, L., Cioffi, R., Cocco, O., Coffetti, D., Colangelo, F., Coppola, B., Corinaldesi, V., Crea, F., ... Yang, F. (2018). Binders alternative to Portland cement and waste management for sustainable construction—part 1. *Journal of Applied Biomaterials & Functional Materials, 16*(3), 186–202. https://doi.org/10.1177/2280800018782845

Dunuweera, S. P., & Rajapakse, R. M. G. (2018). Cement types, composition, uses and advantages of nanocement, environmental impact on cement production, and possible solutions. *Advances in Materials Science and Engineering, 1*, 1–11. https://doi.org/10.1155/2018/4158682

Dyer, T. D. (2014). Glass recycling. In E. Worrell, M.A. Reuter (Eds.), *Handbook of recycling: state-of-the-art for practitioners, analysts, and scientist* (pp. 191–209). https://doi.org/10.1016/B978-0-12-396459-5.00014-3

Fayomi, G. U., Mini, S. E., Fayomi, O. S. I., & Ayoola, A. A. (2019). Perspectives on environmental CO_2 emission and energy factor in Cement Industry. *IOP Conference Series: Earth and Environmental Science, 331*(1), 012035. https://doi.org/10.1088/1755-1315/331/1/012035

FEVE Recycling Statistics. (2017). *Record collection of glass containers for recycling hits 76% in the EU*. Brussels, Viewed at https://feve.org/record-collection-of-glass-containers-for-recycling-hits-76-in-the-eu/

Faroug, F., Szwabowski, J., & Wild, S. (1999). Influence of superplasticizers on workability of concrete. *Journal of Materials in Civil Engineering, 11*(2), 1–5. https://doi.org/10.1061/(ASCE)0899-1561(1999)11:2(151)

Gerges, N. N., Issa, C. A., Fawaz, S. A., Jabbour, J., Jreige, J., & Yacoub, A. (2018). Recycled glass concrete: coarse and fine aggregates. *European Journal of Engineering and Technology Research, 3*(1), 1–9. https://doi.org/10.24018/ejers.2018.3.1.533

Geueke, B., Groh, K., & Munchke, J. (2018). Food packaging in the circular economy: Overview of chemical safety aspects for commonly used materials. *Journal of Cleaner Production, 193*(1), 491–505. https://doi.org/10.1016/j.jclepro.2018.05.005

Harrison, E., Berenjian, A., & Seifan, M. (2020). Recycling of waste glass aggregate in cement-based materials. *Environmental Science and Ecotechnology, 4*(1), 10064. https://doi.org/10.1016/j.ese.2020.100064

Henson, K. (2011). *The procurement and use of sustainable concrete on the Olympic Park* (pp. 1–12). Learning Legacy. https://webarchive.nationalarchives.gov.uk/20130403014421; http://learninglegacy.independent.gov.uk/publications/the-procurement-and-use-of-sustainable-concrete-on-the-o.php

Henrist, C., Mathier, J.-P., Vogels, C., Rulmont, A., & Cloots, R. (2003). Morphological study of magnesium hydroxide nanoparticles precipitated in dilute aqueous solution. *Journal of Crystal Growth., 249*(1–2), 321–330. https://doi.org/10.1016/S0022-0248(02)02068-7

Hoang, N.-D., & Pham, A.-D. (2016). Estimating concrete workability based on slump test with least squares support vector regression. *Journal of Construction of Engineering, 1*, 1–8. https://doi.org/10.1155/2016/5089683

Industry Edge Pty Ltd., & Equilibrium OMG Pty Ltd. (2017). *Paper packaging, glass containers, steel cans and aluminium packaging* (pp. 1–13). National Recycling and Recovery Surveys (NRRS). https://documents.packagingcovenant.org.au/public-documents/National%20Recycling%20and%20Recovery%20Survey%20(NRRS)%202015-16%20for%20Paper%20Packaging,%20Glass%20Containers,%20Steel%20Cans%20and%20Aluminum%20Packaging

Islam, G. M. S., Rahman, M. H., & Kazi, N. (2017). Waste glass powder as partial replacement of cement for sustainable concrete practice. *International Journal of Sustainable Built Environment, 6*(1), 37–44. https://doi.org/10.1016/j.ijsbe.2016.10.005

Ismail, Z. Z., & Al-Hashmi, E. A. (2009). Recycling of waste glass as a partial replacement for fine aggregate in concrete. *Waste Management, 29*(2), 655–659. https://doi.org/10.1016/j.wasman.2008.08.012

Kaza, S., & Yao, L. (2018). At a glance: A global picture of solid waste management. In: S. Kaza, L. Yao, P. Bhada-Tata, F.V. Woerden (eds.) *What a waste 2.0: a global snapshot of solid waste management to 2050* (pp. 17–38). https://doi.org/10.1596/978-1-4648-1329-0_ch2

Kanojia, A., & Jain, S. K. (2017). Performance of coconut shell as coarse aggregate in concrete. *Construction and Building Materials, 140*(1), 150–156. https://doi.org/10.1016/j.conbuildmat.2017.02.066

Kou, S. C., & Poon, C. S. (2009). Properties of self-compacting concrete prepared with recycled glass aggregate. *Cement and Concrete Composites, 31*(2), 107–113. https://doi.org/10.1016/j.cemconcomp.2008.12.002

Kumar, L., Li, W. Z., Vannoy, C. H., Leblanc, R. M., & Wang, D. Z. (2009). Synthesis, characterization and optical properties of Mg(OH)2 micro-/nanostructure and its conversion to MgO. *Ceramic International, 32*(8), 3355–3364. https://doi.org/10.1016/j.ceramint.2009.05.035

Li, Y., Wang, J., & Xu, Z. (2016). Design optimization of a concrete face rock-fill dam by using genetic algorithm. *Mathematical Problems in Engineering, 1*, 1–11. https://doi.org/10.1155/2016/4971048

Malhotra, V. M. (1981). Effect of repeated dosages of superplasticizers on slump, strength and freeze-thaw resistance of concrete. *Materiaux Et Construction, 14*(1), 79–89. https://doi.org/10.1007/BF02547642

Matschei, T., Lothenbach, B., & Glasser, F. P. (2007). The role of calcium carbonate in cement hydration. *Cement and Concrete Research, 37*(4), 551–448. https://doi.org/10.1016/j.cemconres.2006.10.013

Matias, D., Brito, J. D., Rosa, A., & Pedro, D. (2014). Durability of concrete with recycled coarse aggregates: Influence of superplasticizers. *Journal of Materials in Civil Engineering, 26*(7), 1–5. https://doi.org/10.1061/(ASCE)MT.1943-5533.0000961

Meldrum-Hanna, C., Davies, A., & Richards, D. (2017). *Recycling companies stockpiling thousands of tonnes of Glass as cheap imports leave market in crisis.* Four Corners, ABC News. Viewed at https://www.abc.net.au/news/2017-08-07/recycling-companies-forced-to-stockpile-glass-industry-crisis/8778088

Mohamed, M., Yusup, S., & Maitra, S. (2012). Decomposition study of calcium carbonate in cockle shell. *Journal of Engineering Science and Technology, 7*(1), 1–10. https://jestec.taylors.edu.my/Vol%207%20Issue%201%20February%2012/Vol_7_1_001_010_SUZANA%20YOUSUF.pdf

Mohammed, M. S., Mohamed, S. A., & Megat Johari, M. Z. (2016). Influence of superplasticizer compatibility on the setting time, strength and stiffening characteristics of concrete. *Advances in Applied Science, 1*(2), 30–36. https://doi.org/10.11648/j.aas.20160102.12

Nafisa, T. (2020). *Use of waste glass as aggregate and cement replacement in concrete.* PhD Thesis, James Cook University, pp. 1–198. https://doi.org/10.25903/nft4%2D8q67

Naqi, A., & Jang, J. G. (2019). Recent progress in green cement technology utilizing low-carbon emission fuels and raw materials: a review. *Sustainability, 11*(2), 537. https://doi.org/10.3390/su11020537

Ogundairo, T. O., Adegoke, D. D., Akinwumi, I. I., & Olofinnade, O. M. (2019). Sustainable use of recycled waste glass as an alternative material for building construction—A review. *IOP Conference Series: Materials Science and Engineering., 640*(1), 012073. https://doi.org/10.1088/1757-899X/640/1/012073

Oztas, A., Pala, M., Ozbay, E., & KAnca, E., Caglar, N., Bhatti, M.A. (2006). Predicting the compressive strength and slump of high strength concrete using neural network. *Construction and Building Materials., 20*(9), 769–775. https://doi.org/10.1016/j.conbuildmat.2005.01.054

Palmquist, S. (2003). *Compressive behaviour of concrete with recycled aggregates.* PhD Thesis, Tufts, University

Park, S. B., Lee, B. C., & Kim, J. H. (2004) Studies on mechanical properties of concrete containing waste glass aggregate. *Cement and Concrete Research, 34*(120, 2181–2189. https://doi.org/10.1016/j.cemconres.2004.02.006

Qin, D., Hu, Y., & Li, X. (2021). Waste glass utilization in cement-based materials for sustainable construction: A review. *Crystals, 11*(6), 710. https://doi.org/10.3390/cryst11060710

Rachael, O., Matthew, A., Popoola, Oyinade, C., Ogundare, & Damilola, A. (2019) Partial replacement of coarse aggregate with waste glass in concrete production. *IOSR Journal of Mechanical and Civil Engineering, 16*(3), 63–69. http://www.iosrjournals.org/iosr-jmce/papers/vol16-issue3/Series-4/J1603046369.pdf

Rashad, A. M. (2014). Recycled waste glass as fine aggregate replacement in cementitious materials based on Portland cement. *Construction and Building Materials, 72*(1), 340–357. https://doi.org/10.1016/j.conbuildmat.2014.08.092

Rashad, A. M. (2015). Recycled cathode ray tube and liquid crystal display glass as fine aggregate replacement in cementitious materials. *Construction and Building Materials, 93*(1), 1236–1428. https://doi.org/10.1016/j.conbuildmat.2015.05.004

Salem, M., Alsadey, S., & Johari, M. (2016) Effect of superplasticizer dosage on workability and strength characteristics of concrete. *IOSR Journal of Mechanical and Civil Engineering, 13*(4), 153–158. http://iosrjournals.org/iosr-jmce/papers/vol13-issue4/Version-7/T130407153158.pdf

Seifan, M., Mendoza, S., & Berenjian, A. (2020). Mechanical properties and durability performance of fly ash based mortar containing nano- and micro-silica additives. *Construction and Building Materials, 252*(1), 119121. https://doi.org/10.1016/j.conbuildmat.2020.119121

Seifan, M., Samani, A. K., & Berenjian, A. (2016). Bioconcrete: Next generation of self-healing concrete. *Applied Microbiology and Biotechnology, 100*(1), 2591–2602. https://doi.org/10.1007/s00253-016-7316-z

Shayan, A., & Xu, A. (2004). Value-added utilization of waste glass in concrete. *Cement and Concrete Research, 34*(1), 81–89. https://doi.org/10.1016/S0008-8846(03)00251-5

Shubbar, A. A., Jafer, H., Abdulredha, M., Al-Khafaji, Z. S., Nasr, M. S., Al Masoodi, Z., & Sadique, M. (2020). Properties of cement mortar incorporated high volume fraction of GGBFS and CKD from 1 day to 550 days. *Journal of Building Engineering, 30*(1), 101327. https://doi.org/10.1016/j.jobe.2020.101327

Singh, S., Srivastava, V., & Agarwal, V. C. (2015) Glass waste in concrete: Effect on workability and compressive strength. *International Journal of Innovative Research in Science, Engineering and Technology, 4*(9), 8142–8150. http://ijirset.com/upload/2015/september/18_Glass.pdf

Spring, S. (2008). *Cement. Ullmann's Encyclopedia of Industrial Chemistry*. Wiley. https://doi.org/10.1002/14356007.a05_489.pub2

Terro, M. J. (2006). Properties of concrete made with recycled crushed Glass at elevated temperatures. *Building and Environment, 41*(5), 633–639. https://doi.org/10.1016/j.buildenv.2005.02.018

Thoo, F. V. W., Zainuddin, N., Matori, K. A., & Abdullah, S. A. (2013). Studies on the potential of waste soda lime silica glass in glass ionomer cement production. *Advances in Materials Science and Engineering, 1*, 1–6. https://doi.org/10.1155/2013/395012

Tittarelli, F., Giousue, C., & Mobili, A. (2018). Recycled glass as aggregate for architectural mortars. *International Journal of Concrete Structures and Materials, 57*(1), 1–20. https://doi.org/10.1186/s40069-018-0290-3

Topcu, I. B., & Canbaz, M. (2004). Properties of concrete containing waste glass. *Cement and Concrete Research, 34*(2), 267–274. https://doi.org/10.1016/j.cemconres.2003.07.003

Topcu, I. B., Boga, A. R., & BIlir, T. (2008). Alkali–silica reactions of mortars produced by using waste glass as fine aggregate and admixtures such as fly ash and Li2CO3. *Waste Management, 28*(5), 878–884. https://doi.org/10.1016/j.wasman.2007.04.005

Wu, X.-F., Hu, G.-S., Wang, B.-B., & Yang, Y.-F. (2008). Synthesis and characterization of superfine magnesium hydroxide with monodispersity. *Journal of Crystal Growth, 310*(2), 457–461. https://doi.org/10.1016/j.jcrysgro.2007.10.025

Yaacob, I. I., Ali, M. Y., Sopyan, L., & Hashmi, S. (2015). Effect of calcium carbonate replacement on workability and mechanical strength of Portland cement concrete. *Advanced Materials Research, 1115*(1), 137–131. https://doi.org/10.4028/www.scientific.net/AMR.1115.137

Glass Waste as Fine Aggregate Filler Replacement in Concrete Addition of Superplasticizer

Chin Mei Yun, Md Rezaur Rahman, Kuok King Kuok, Amelia Chai Pei Sze, Kenneth Jong Kai Zhiing, and Muhammad Khusairy Bin Bakri

Abstract Glass has been made and used in various ways, such as bulbs, bottles, and cathode-ray tubes, all of which have a limited lifespan and are disposed of in landfills, causing multiple environmental concerns. Due to the physical and chemical similarities between glass and sand, glass waste can be used in concrete mixing by substituting glass waste for partial fine aggregates of sand. This study aims to determine the best percentage of glass waste as fine aggregates in concrete, both with and without a superplasticizer. The study looks at how varying percentages of sand substituted with glass trash affect the compressive strength of concrete. The first step of the experiment involved replacing fine aggregates with glass trash at a ratio of 15, 20, 25, and 0% as the control group. The second step entails combining the optimal percentage replacement from the first portion with superplasticizer at varying dosages of 0.8, 1, and 1.2% of the total cementitious weight. After 7 and 28 days of curing, the samples were subjected to a compressive strength test and characterization. According to the findings, samples with replacement percentages of 15, 20, and 25% showed better compressive strength than the control group sample, supporting the characterization result. In the case of the superplasticizer admixture, the compressive strength has risen considerably for each dose sample compared to the control sample. In conclusion, the research reveals encouraging findings for replacing sand with glass in concrete up to 20% by weight of sand, with the addition of a superplasticizer if higher strength is needed.

Keywords Concrete · Aggregates · Glass · Waste · Strength

C. M. Yun · K. K. Kuok · A. C. P. Sze · K. J. K. Zhiing
Faculty of Engineering, Computing and Science, Swinburne University of Technology Sarawak Campus, Jalan Simpang Tiga, 93400 Kuching, Sarawak, Malaysia

M. R. Rahman (✉) · M. K. B. Bakri
Faculty of Engineering, Universiti Malaysia Sarawak, Jalan Datuk Mohamad Musa, 94300 Kota Samarahan, Sarawak, Malaysia
e-mail: rmrezaur@unimas.my

M. R. Rahman et al. (eds.), *Waste Materials in Advanced Sustainable Concrete*, Engineering Materials, https://doi.org/10.1007/978-3-030-98812-8_3

1 Introduction

Concrete is a widely used material in the construction industry, especially in this era of globalization. Therefore it has caused a high demand for concrete production (Biernacki et al., 2017). According to Ismail et al. (2013), there is a risk of a scarcity of natural aggregates used in concrete, such as sand and rocks, due to the daily increase in concrete manufacturing. As a result, demand for alternative aggregates is crucial to avoid the depletion of natural aggregates (Ismail et al., 2013; Makul et al., 2021; Mavroulidou, 2017). Because of its physical and chemical similarities to sand, glass is one of the ideal materials for use in concrete, according to previous studies (Islam et al., 2017; Kadir et al., 2016; Malek et al., 2020; Olofinnade et al., 2018; Tayeh et al., 2018). As a result, glass debris from discarded bottles may be recovered and utilized as a concrete aggregate.

According to the United Nations, glass makes about 7%, which is 14 million tons of global trash, therefore using glass waste in concrete manufacturing might have a substantial environmental impact (Jani & Hogland, 2014). Furthermore, owing to the expensive recycling procedure, glass has one of the lowest recycling rates compared to other trash, resulting in it being thrown into landfills (Hopewell et al., 2009; Johari et al., 2014; Needhidasan et al., 2014; Rasmussen et al., 2005). Glass is a recyclable material that may be recycled indefinitely without losing any of its chemical characteristics. As a result, a study of the characteristics of glass waste as a partial substitute for fine aggregates in concrete with and without the use of a superplasticizer was conducted (Ibrahim, 2017).

2 Optimum Waste Glass Replacement

According to Ismail and Al-Hashmi (2009), research was done with a combination of cement (380 kg/m^3), sand (715 kg/m^3), gravel (1020 kg/m^3), and water (201 kg/m^3) to substitute 0, 10, 15, and 20% of fine aggregates with waste glass. Each batch of concrete with a variable proportion of waste glass replacement was allowed to cure for 3, 7, 14, and 28 days. As illustrated in Table 1, the concrete with 20% waste glass replacement at 28 days had the highest compressive strength of 25.9 MPa.

Table 1 Compressive strength of waste glass mixture (Ismail & Al-Hashmi, 2009)

Curing ages (days)	3	7	14	28
Percentage of waste glass (%)	Compressive strength (MPa)			
0	26.9	31.5	43.8	44.0
10	29.1	34.6	39.1	40.3
15	28.9	32.0	38.3	42.0
20	27.6	31.7	38.0	45.9

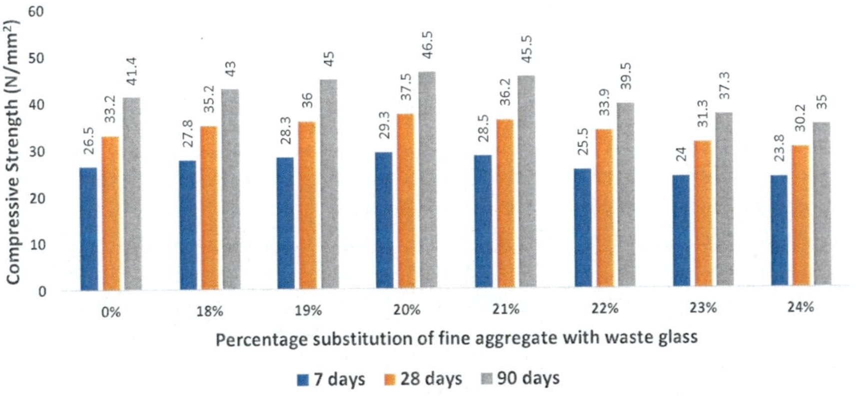

Fig. 1 Compressive strength of concrete with different percentage waste glass mixture at 7, 28, and 90 days (Bisht and Ramana, 2018)

In comparison to the control group, the best outcome had a 4.23% improvement in strength. According to Ismail and Al-Hashmi (2009), waste glass concrete cured for 14 days had lower compressive strength than the control group, which might be attributed to a decrease in adhesive strength between the cement paste and the waste glass surface area over that time. As a result, the pozzolanic impact of waste glass in concrete is more noticeable at a 28-day curing age (Du & Tan, 2014; Gimenez-Carbo et al., 2021; Kou & Xing, 2012; Mirzahosseini & Riding, 2014; Tariq et al., 2020). As a result, fine waste glass in place of fine particles in concrete might be favourable if the glass geometry is less heterogeneous (Abdallah & Fan, 2014; Al-Bawi et al., 2017; Sales et al., 2017).

According to Bisht and Ramana (2018), the experiment of waste glass used as a fine aggregates replacement in concrete done by them varies at portions of 0, 18, 19, 20, 21, 22, 23, and 24% with the curing period of 7, 28 and 90 days. Three cubes (100 × 100 × 100 mm) from each batch were subjected to a compressive strength test at a rate of 140 kg/cm^2/min. The results revealed that compressive strength improves as the proportion of glass replaced increases up to 20%, after which the strength declines steadily after 20% replacement, as shown in Fig. 1. Because compressive strength improved with age, 20% waste glass replacement with 90 days of curing had the best compressive strength of all specimens (Al-Swaidani et al., 2017; He et al., 2019; Islam et al., 2017; Kumar et al., 2021).

3 Effect of Superplasticizer in Concrete

Sulphonated naphthalene formaldehyde (SNF), sulphonated melamine formaldehyde (SMF), polycarboxylic ether (PCE), and lignosulphonate (LS) were used in compressive strength tests conducted by Manomi et al. (2018) on different percentages of fly ash content in concrete with different types of superplasticizers. The compressive strength of 0 and 15% fly ash content concrete was more significant than the control mix, whereas 25 and 35% fly ash content concrete had a lower strength than the control mix, as shown in Fig. 2 and Table 2. Given that the fly ash concentration of the concrete does not exceed 15%, it was found that the SNF superplasticizer enhances the concrete's strength.

At all three curing periods (1, 7, and 28 days), Benaicha et al. (2019) found that concrete containing 0.3% superplasticizer had higher compressive strength than concrete with no superplasticizer. On the other hand, the compressive strength diminishes when the superplasticizer dose goes from 0.3 to 0.7%, when the strength falls

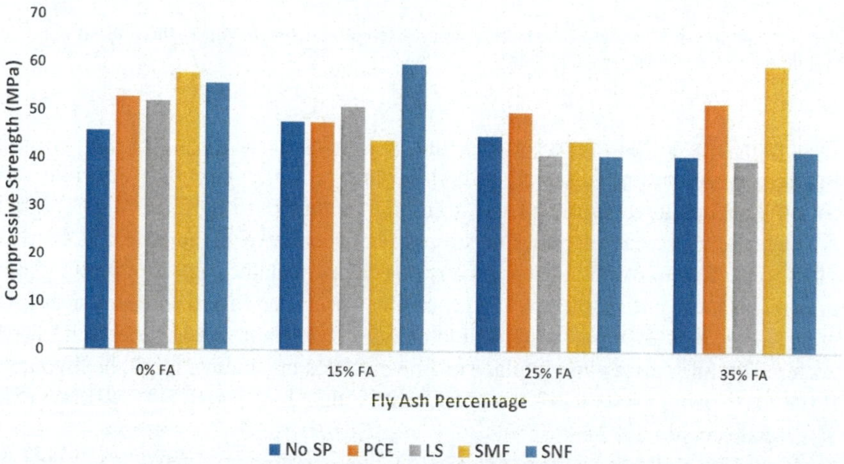

Fig. 2 Compressive strength of concrete with different types of superplasticizers and fly ash content (Manomi et al., 2018)

Table 2 Compressive strength of concrete with different dosages of superplasticizer (Benaicha et al., 2019)

Mixtures SCC's	Compressive strength (MPa)		
	1 day	7 days	28 days
N	29.4	33.3	50.8
SCC-SP1	45.2	60.8	73.48
SCC-SP2	40.64	55.48	71.44
SCC-SP3	35.3	46.28	65.2
SCC-SP4	32.24	41.5	59.88
SCC-SP5	29.24	37.84	53.24
SCC-SP6	26.64	33.56	42.24
SCC-SP7	18.88	27.24	38.7
SCC-SP8	15.04	20.76	29.44

below the control groups. As a result, if the proper amount of superplasticizer is used, the concrete with superplasticizer has a higher compressive strength than the concrete without superplasticizer (Alaka et al., 2016; Mohammed et al., 2016).

According to the findings, using waste glass as a fine aggregate substitute in concrete decreased environmental issues and enhanced the strength of the concrete (Abdallah & Fan, 2014; Meddah, 2019; Rashad, 2014). It has also been proved that replacing 20% of the fine aggregate with waste glass gives the concrete the best compressive strength. In terms of superplasticizers, it was discovered that adding the appropriate quantity of superplasticizer to concrete can enhance its strength (Sathyan et al., 2017; Xun et al., 2020). However, using too much superplasticizer reduces the strength of the concrete, which may eventually be lower than the strength of concrete without it (Alsadey, 2015; Farhan et al., 2020). Nonetheless, research on substituting fine particles in concrete with waste glass and adding a superplasticizer has yet to be completed. Therefore, the experimental investigation was completed, and the findings were summarized in the following report.

4 Methodology

Two stages of experiments were conducted in this study to determine the workability and compressive strength of the concrete. The first step entails replacing 15, 20, and 25% sand with glass waste. In contrast, the second phase entails expanding on the first phase by adding various dosages of superplasticizer to the optimal waste glass percentage replacement from the first phase.

4.1 Materials

Glass waste (Heineken glass bottles) was gathered from a nearby restaurant. A sand crusher machine was used to break the glass bottles into particles smaller than 2 mm. Cahaya Mata Sarawak (CMS) Berhad, Sarawak's largest cement manufacturer and supplier, provided the Portland cement utilized in this experiment. The coarse materials used in this study were 100% crushed rocks purchased from a local source and sieved into four sizes. 12, 24, 24, and 40% of the crushed rocks utilized had particle sizes of 20, 14, 10, and 5 mm, respectively. In this experiment, fine aggregate made from river sand purchased locally and crushed waste glass were utilized as fine aggregates with a maximum particle size of 2 mm. The percentage of the two fine components in the mix varies. An imported superplasticizer was being used in this experiment. In this experiment, the superplasticizer dosages were 0.8, 1, and 1.2% of the total cementitious weight. The following is a list of the superplasticizers used, Mighty 150, manufactured by Kao Industrial (Thailand) Co. Ltd. Naphthalene sulfonate formaldehyde superplasticizer contained mostly naphthalene sulfonic acid, polymer with formaldehyde, sodium salt, and water.

Table 3 Mix proportion for phase 1

Specimen	Cement (kg/m^3)	Gravel (kg/m^3)	Sand (kg/m^3)	Percentage of waste Glass (%)	Water (kg/m^3)	Superplasticizer (% of total cementitious weight)
OPC	400	1100	730.0	0	208	0
WG15	400	1100	620.5	15	208	0
WG20	400	1100	584.0	20	208	0
WG25	400	1100	547.5	25	208	0

Table 4 Mix proportion for phase 2

Specimen	Cement (kg/m^3)	Gravel (kg/m^3)	Sand (kg/m^3)	Percentage of waste glass (%)	Water (kg/m^3)	Superplasticizer (% of total cementitious weight)
WG20SP0.8	400	1100	584.0	20	184	0.8
WG20SP1.0	400	1100	584.0	20	184	1
WG20SP1.2	400	1100	584.0	20	184	1.2

4.2 *Concrete Mix Proportion*

In this study, three different types of concrete mixtures were used. The control group (0%) comprised of Portland cement (400 kg/m^3), sand (730 kg/m^3), gravel (1100 kg/m^3), and water (208 kg/m^3), yielding a water-concrete ratio of 0.52. The second concrete mix used discarded glass aggregates to replace 15, 20, and 25% of the sand in the mix, along with the same quantity of cement, gravel, and water. The third concrete mix had the same quantity of cement and gravel as phase 1 but with various superplasticizer dosages at 0.8, 1, and 1.2% of the total cementitious weight. Due to the superplasticizer functioning as a water reduction agent, the water content of the third mix was lowered to 184 kg/m^3 compared to the previous two kinds, resulting in a water-concrete ratio of 0.46. Tables 3 and 4 are the phase 1 and phase 2 mix proportions.

4.3 *Casting and Curing of Concrete*

For each batch, three cubes measuring 100 mm × 100 mm × 100 mm were cast. Before being placed in the water curing tank, the concrete cubes were kept in the mold for 24 h. The cubes were left to cure for 7 and 28 days, respectively. For this research project, a total of 42 cubes were produced.

4.4 Slump Test

A slump test was performed for each batch in Phase 1 and Phase 2 before casting to assess the workability of each kind of combination using a slump cone, base plate, tamping rod, and ruler. As a guide, the ASTM C143/C143M-20 (2020) standard was employed.

4.5 Compressive Strength Test of Concrete

After 7 and 28 days of curing, each cube was tested for compressive strength. For each batch, the average results were obtained from the three cubes cast. The compressive strength test was performed using Unit Test Autocon 2000 in compliance with BS EN 12,390-3 (2019). During the compressive strength test, the load was delivered to the mortar specimens gradually at 0.05 ± 0.01 kN/s across an area of 100 mm × 100 mm until the samples failed. The results were averaged and reported from mortar specimens from the same batch.

4.6 Scanning Electron Microscopy (SEM) with Energy Dispersive Spectroscopy (EDX/EDS)

Scanning electron microscopy (SEM) and energy dispersive spectroscopy (EDX/EDS) were performed under ASTM standards, respectively. The samples were examined at magnifications of 100×, 250×, 500×, and 10,000×. The elemental composition percentages of the samples are scanned and analyzed by automated software. The EDX/EDS was performed many times on each sample at various stages, with the most representative results chosen. For the materials' SEM and EDX/EDS examinations, Hitachi Ltd., Tokyo, Japan, utilized a Hitachi TM4000Plus Tabletop Microscope with a Quantax75TM Series Energy Dispersive X-Ray Spectrometer (Hitachi Ltd., Tokyo, Japan).

4.7 Fourier Transform Infrared (FTIR) Spectroscopy

FTIR materials were analyzed using Fourier-transform infrared spectroscopy (IRAffinity-1, Shimadzu Corporation, Kyoto, Japan). Fourier-transform infrared spectroscopy was performed for qualitative and quantitative examination according to ASTM standards. Each sample had its spectra scanned in the wavenumber range of 4000–400 cm^{-1}. The samples' infrared spectrum transmittance and absorption were used to create a unique chemical fingerprint spectrum using Fourier-transform infrared spectroscopy. For each sample, the test was performed many times, and the most representative findings were chosen.

5 Results and Discussions

5.1 Workability Properties

Slump tests are used to assess the concrete's workability, with a target range of 75–100 mm. Figure 3 depicts the slump test findings from phases 1 and 2 as blue and red, respectively. According to Fig. 3, all phase 1 concrete mixes have attained the necessary workability with a slump of 75–100 mm. The workability of each batch, on the other hand, displays a wide range of slump values. One probable explanation is that the concrete mixture had an uneven moisture content when cast because the laboratory was not entirely appropriate for casting concrete (Chakravarthi, 2014). Other than that, materials used, such as Portland cement and sand, may not have been appropriately stored, affecting the materials' characteristics (Dunuweera & Rajapakse, 2018). In comparison to phase 1, the slump test findings from phase 2 were low. One probable explanation is that owing to the presence of superplasticizer in phase 2, the amount of water added to the mixes was lower than in Phase 1. As a result, phase 2's workability for concrete mixes was lower than phase 1.

Singh et al. (2015) show that fine aggregate with superplasticizer and fine aggregate with different percentages have different effects on the concrete's workability. As the fine glass replacement of a fixed superplasticizer concrete increases, workability increases. This increase was due to the increasing content of waste glass that

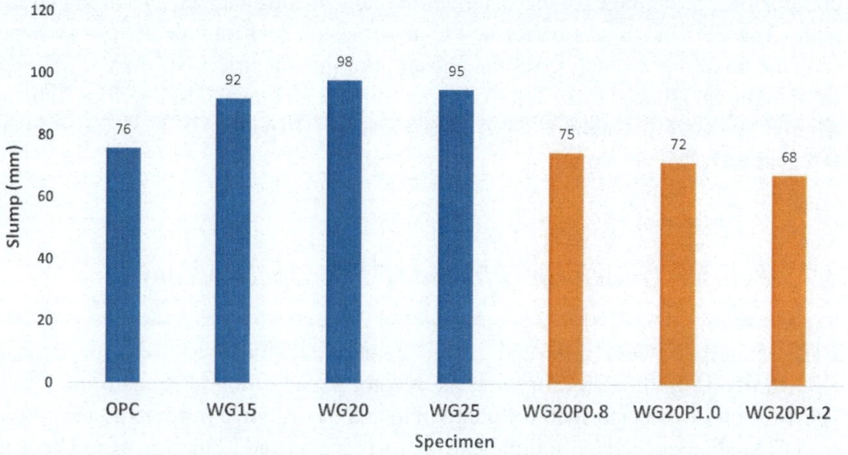

Fig. 3 Slump for phases 1 and 2

is hydrophobic. The result causes more water availability as compared to conventional concrete (Singh et al., 2015). Increased workability may result from waste glass's non-absorbent quality compared to natural sand (Khalil & Al-Obeidy, 2018). However, the reduction in a slump may be attributable to the loss of sufficient water from samples during fast transit owing to the increased voids created by comparable waste glass components, rather than natural sand being heavier than water.

As the proportion of glass grows, Ekwulo and Eme (2018) found that both processes' concrete workability decreases after it reaches its stability threshold. The difference in density between the glass and the aggregates is likely to cause the reduced workability of the concrete generated (Ekwulo & Eme, 2018). Ekwulo and Eme (2018) results also revealed that the workability of glass-granite concrete is superior to glass-sand concrete for percentage replacements of 5–25%, beyond which the opposite is true. There is a significant drop in concrete workability when granite is replaced with glass at 30%. The decline in workability in the glass-sand scenario, on the other hand, follows a constant trend as the percentage replacement increases.

5.2 *Comprehensive Strength Properties*

Figures 4 and 5 show the concrete's 7 and 28-day compressive strength from phases 1 and 2. Figure 4 shows that WG20 has the maximum compressive strength of 32.66 MPa after 7 days of curing, despite WG15's strength reduction compared to OPC. However, when the waste glass content grows, the compressive strength of 28-day concrete for phase 1 increases. To obtain better strength characteristics, WG15 required a longer curing period. Many researchers suggested a longer curing

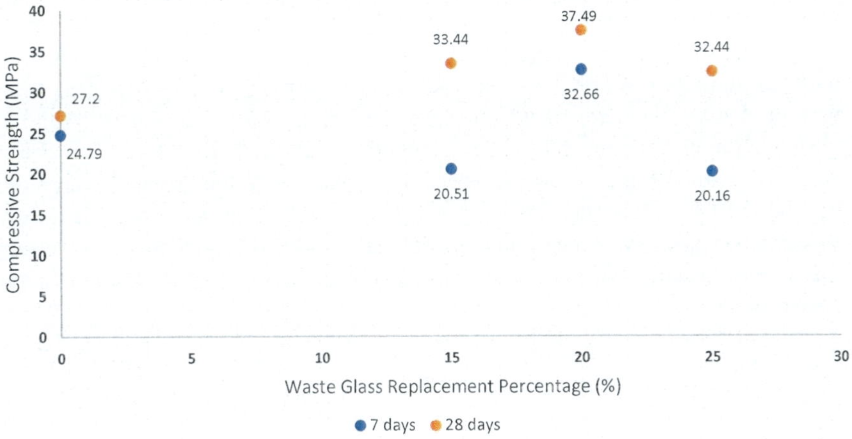

Fig. 4 Compressive strength for Phase 1

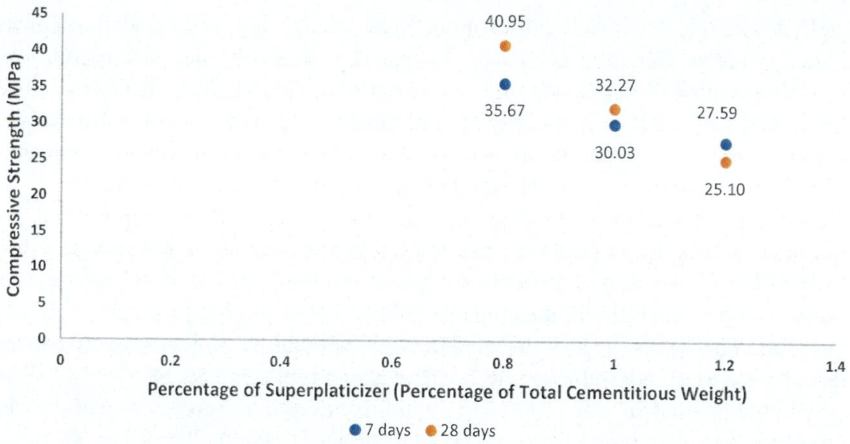

Fig. 5 Compressive strength for Phase 2

period to allow cross-linkage between materials to bind and hardened concrete (Chin et al. ; Pacewska & Wilinska, 2020; Safiuddin et al., 2018; Saraswathy et al., 2017; Thomas, 2007). Furthermore, compressive strength began to drop after 25% waste glass replacement; thus, WG20 has the optimal waste glass replacement for concrete strength of 7 and 28 curing days with 20% waste glass content.

According to the Phase 2 experimental results in Fig. 5, the compressive strength falls as the superplasticizer dosage rises. With the inclusion of 0.8% total cementitious weights of superplasticizer and 20% fine waste glass content, WG20SP0.8 has the maximum compressive strength for both 7 and 28 curing days at 35.67 and 40.95 MPa, respectively. Aside from that, WG20SP0.8 has a strength improvement of roughly 3 MPa above WG20 without using a superplasticizer.

According to Ekwulo and Eme (2018), the best replacement for granite is with glass as a fine aggregate component. Glass may also be utilized as a replacement for sand up to a 35% replacement level, according to their findings. As compared with our findings, it is noted that the fine aggregate can only reach up to 20%, which may be due to the factor of different types of glass composition used. Although glass-sand aggregate generated concrete with slightly higher strength than control concrete (with 0% glass), coarse aggregate replacement should not be promoted beyond a 10% replacement level since concrete with lesser strength was created. When the compressive strength of concrete was examined at both optimal levels, the fine aggregate generated substantially greater compressive strength concrete than the coarse aggregate.

5.3 Morphological Properties

Figure 6a–d show SEM images of the WG20 (7 days) at magnifications of 100x, 250x, 500x, and 1000x, respectively. The surface sample's topology was flakily mixed with an unreacted structure that was quite weak, as seen in Fig. 6. The low hydraulic reactivity and the lack of chemical or physical contact create a soft interlock that prevents it from combining with other materials (Balasubramanian et al., 2021). According to other studies, the compact microstructure was achieved using the right paste mixture, soundness, and hydraulic reactivity (Chin et al., 2020a, 2020b; Balasubramanian et al., 2021). It also noted that due to the glass lower density than the other mineral used in the paste or concrete, it could easily rise on the concrete's surface, which caused lower accumulation in the middle or bottom of the concrete if no plasticizer were used (Khan et al., 2014). Small holes, pores, or bubbles are also noticed in the structure of the concrete. It might be due to the release of trapped air or bubbles during mixing and molding. According to research, trapped air or bubbles bring both good and bad benefits to the concrete (Zeng et al., 2020). Bigger or many air bubbles reduce the strength of the concrete, while small and fewer bubbles bring extra strength to the concrete. It also depends on the locality and spread area of the bubbles or pores.

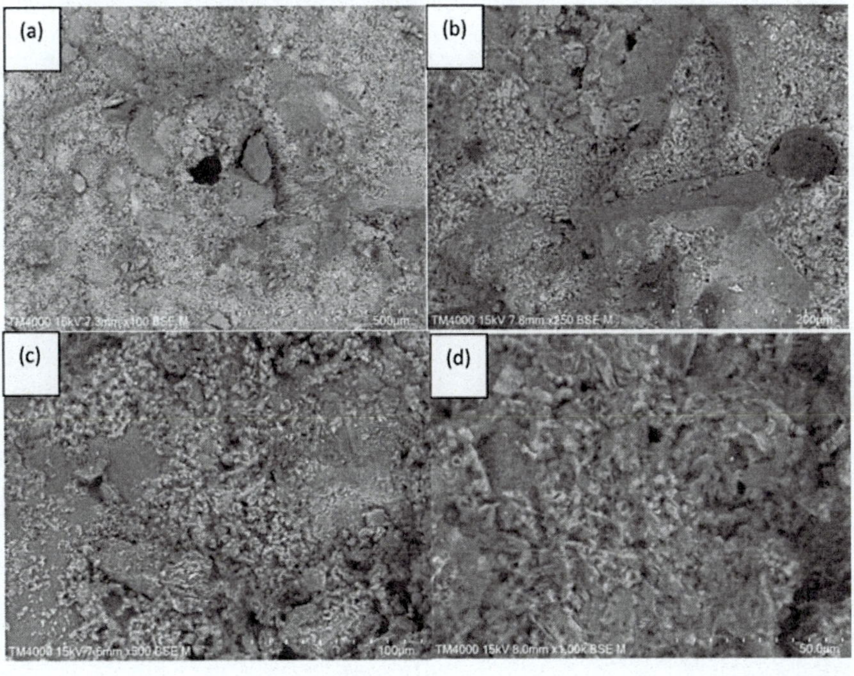

Fig. 6 SEM images for WG20 (7 days) at magnification of **a** 100×, **b** 250×, **c** 500×, and **d** 1000×

5.4 Elemental Properties

The EDS/EDX plot and element contents for WG20 (7 days) are shown in Table 5 and Fig. 7. Carbon (C), oxygen (O), and calcium (Ca) were found in the sample, as well as silicon (Si), aluminum (Al), iron (Fe), and sulfur (S). Table 5 shows that oxygen has the highest mass proportion, followed by calcium and carbon. As the paste and hydraulic reactivity are mixed, a chemical reaction occurs, resulting in calcium carbonate, $CaCO_3$. $CaCO_3$ improves early strength by acting as an accelerator and having a high rate of hydration, causing the concrete to solidify more quickly (Matschei et al., 2007; Yaacob et al., 2015). SiO_2, Al_2O_3, and Fe_2O_3 were other chemical reactions that showed the presence of glass in the concrete (Islam

Table 5 Element composition for WG20 (7 days)

Element	Mass normal (%)	Atom (%)
Ca	45.25	25.94
O	43.52	62.48
Si	5.66	4.63
C	2.64	5.04
Al	1.16	0.98
Fe	1.15	0.47
S	0.68	0.45
Sum	100.00	100.00

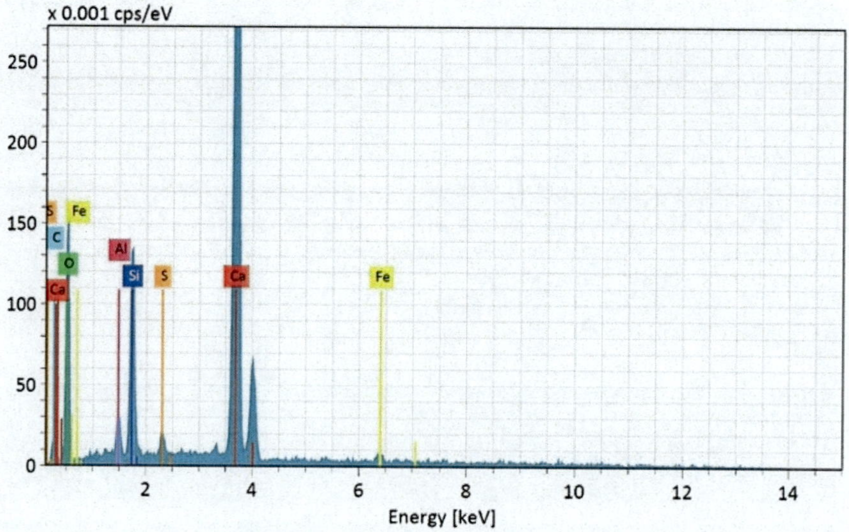

Fig. 7 EDS/EDX graph for WG20 (7 days)

et al., 2017). The amount of SiO_2 is determined by how much fine glass waste is used in the concrete. The more the glass percentage, the more SiO_2 amount is in the concrete (Wang et al., 2020). In addition to the reaction products formed with high ratios of and Ca–Si were favorable in mitigating the concrete alkaline-silica expansion (Cai et al., 2019). It would lead to high strength in the mechanical properties until it reaches the optimum mixture. However, the over-amount of Ca–Si and its expansion may create weak mechanical properties in concrete after it reaches the optimum level.

5.5 *Infrared Spectral Properties*

The FTIR spectra for WG20 (7 days) are shown in Fig. 8. O–H bond, which was owing to hydraulic of water bond and moisture, produced a small broad peak at the 3200–3500 cm^{-1} band. Si–O–Si and Si–OH stretching was found in the peak bands for 1697.36, 1521.84, 578.72, and 536.21 cm^{-1}. It should be noted that polymerization of Si–OH to Si–O–Si occurs during cement setting (Thoo et al., 2013). According to other studies, stretching moves first to lower wavenumbers when polymerization decreases and later to higher wavenumbers, presumably suggesting growth in jennite-like structural settings (Yu et al., 2004). The formation of Ca–Si also happened at the peaks of 578.72 and 536.21 cm^{-1}, which is consistent with single silicate chain structures (Yu et al., 2004).

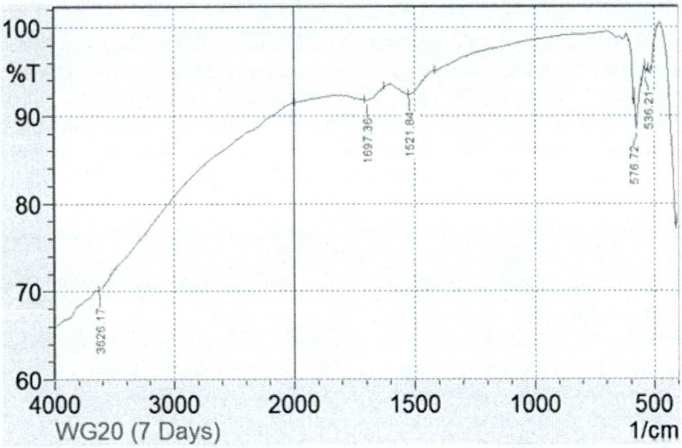

Fig. 8 FTIR for WG20 (7 days)

6 Conclusions

Finally, experimental research was carried out on the characteristics of glass waste as a partial substitute for fine particles in concrete, both with and without using a superplasticizer. The suggested substitution percentage of waste glass as fine aggregates is 20%, with a superplasticizer addition dose of 0.8% of the total cementitious weight. The best result produced with compressive strength of more than 40 MPa can be used in structural applications. This experimental investigation is thought to help society, the environment, and the concrete manufacturing industry. It is advised that the dose of the superplasticizer be investigated further by lowering the dosage to establish the optimal strength of the superplasticizer dosage.

Acknowledgements The authors would like to thank Swinburne University of Technology Sarawak Campus and Universiti Malaysia Sarawak (UNIMAS) for the collaboration.

References

Abdallah, S., & Fan, M. (2014). Characteristics of concrete with waste glass as fine aggregate replacement. *International Journal of Engineering and Technical Research, 2*(6), 11–17. https://www.erpublication.org/published_paper/IJETR021927.pdf

Alaka, H. A., Oyedele, L. O., & Toriola-Coker, O. L. (2016). Effect of excess dosages of superplasticizer on the properties of highly sustainable high-volume fly ash concrete. *International Journal of Sustainable Building Technology and Urban Development, 7*(2), 73–86. https://doi.org/10.1080/2093761X.2016.1167643

Al-Bawi, R. K., Kadhim, I. T., & Al-Kerttani, O. (2017). (2017) Strengths and failure characteristics of self-compacting concrete containing recycled waste glass aggregate. *Advances in Materials Science and Engineering, 1*, 1–12. https://doi.org/10.1155/2017/6829510

Al-Swaidani, A., Soud, A., & Hammami, A. (2017). Improvement of the early-age compressive strength, water permeability, and sulfuric acid resistance of scoria-based mortars/concrete using limestone filler. *Advances in Materials Science and Engineering, 1*, 1–17. https://doi.org/10.1155/2017/8373518

Alsadey, S. (2015). Effect of superplasticizer on fresh and hardened properties of concrete. *Journal of Agricultural Science and Engineering, 1*(2), 70–74.

ASTM C143/C143M-20. (2020). *Standard test method for slump of hydraulic-cement concrete*. ASTM International. https://doi.org/10.1520/C0143_C0143M-20

Balasubramanian, B., Gopala Krishna, G. V. T., Saraswathy, V., & Srinivasan, K. (2021). Experimental investigation on concrete partially replaced with waste glass powder and waste E-plastic. *Construction and Building Materials, 278*(1), 122400. https://doi.org/10.1016/j.conbuildmat.2021.122400

Benaicha, M., Hafidi Alaoui, A., Jalbaud, O., & Burtschell, Y. (2019). Dosage effect of superplasticizer on self-compacting concrete: Correlation between rheology and strength. *Journal of Materials Research and Technology, 8*(2), 2063–2069. https://doi.org/10.1016/j.jmrt.2019.01.015

Bhist, K., & Ramana, P. V. (2018). Sustainable production of concrete containing discarded beverage glass as fine aggregate. *Construction and Building Materials, 177*(1), 116–124. https://doi.org/10.1016/j.conbuildmat.2018.05.119

Biernacki, J. J., Bullard, J. W., Sant, G., Brown, K., Glasser, F. P., Jones, S., Ley, T., Livingston, R., Nicoleau, L., Olek, J., Sanchez, F., Sahsavari, R., Stutzman, P. E., Sobolev, K., & Prater, T.

(2017). Cements in the 21st century: Challenges, perspectives, and opportunities. *Journal of the American Ceramic Society, 100*(7), 2746–2773. https://doi.org/10.1111/jace.14948

BS EN 12390-3. (2019). *Testing hardened concrete. Compressive strength of test specimens*. British Standard Institution.

Cai, Y., Xuan, D., & Poon, C. S. (2019). Effects of nano-SiO2 and glass powder on mitigating alkali-silica reaction of cement glass mortars. *Construction and Building Materials, 201*(1), 295–302. https://doi.org/10.1016/j.conbuildmat.2018.12.186

Chakravarthi, B. A. (2014). *Study on different types of cracks in plain and reinforced concrete*. Master Thesis. Gitam Institute of Technology, pp. 1–121.

Chin, M. Y., Rahman, M. R., Phing, C. Y. W., Chie, A. W. M., & Bakri, M. K. B. (2020). The curing times effect on the strength of ground granulated blast furnace slag (GGBFS) mortar. *Construction and Building Materials, 260*(1), 120622. https://doi.org/10.1016/j.conbuildmat.2020.120622

Chin, M. Y., Rahman, M. R., Kuok, K. K., Chiew, W. Y., & Bakri, M. K. B. (2020) Characterization and impact of curing duration on the compressive strength of coconut shell coarse aggregate in concrete. *BioResources, 16*(3), 6057–6073. https://ojs.cnr.ncsu.edu/index.php/BioRes/article/view/BioRes_16_3_6057_Chin_Curing_Duration_Compressive_Strength_Coconut_Shell

Du, H., & Tan, K. H. (2014). Waste glass powder as cement replacement in concrete. *Journal of Advanced Concrete Technology, 12*(11), 468–477. https://doi.org/10.3151/jact.12.468

Dunuweera, S. P., & Rajapakse, R. M. G. (2018). Cement types, composition, uses and advantages of nanocement, environmental impact on cement production, and possible solutions. *Advances in Materials Science and Engineering, 1*, 1–11. https://doi.org/10.1155/2018/4158682

Ekwulo, E. A., & Eme, D. B. (2018). Comparative study on compressive strength of coarse and fine glass-aggregate concrete. *American Journal of Engineering Research, 2*, 271–277.

Farhan, W. K. M., Bakar, H. B. A., Hilal, N., & Al-Hadithi, A. I. (2020). Effect of silica fume and super-plasticizer on mechanical properties of self-compacting concrete: A review. *IOP Conference Series: Materials Science and Engineering, 978*(1), 1–10. https://doi.org/10.1088/1757-899X/978/1/012052

Gimenez-Carbo, E., Soriano, L., Roig-Flores, M., & Serna, P. (2021). Characterization of glass powder from glass recycling process waste and preliminary testing. *Materials, 14*(11), 2971. https://doi.org/10.3390/ma14112971

He, Z.-H., Zhan, P.-M., Du, S.-G, Liu, B.-J., & Yuan, W.-B. (2019). Creep behavior of concrete containing glass powder. *Composites Part B: Engineering, 166*(1), 13–20. https://doi.org/10.1016/j.compositesb.2018.11.133

Hopewell, J., Dvorak, R., & Kosior, E. (2009). Plastics recycling: Challenges and opportunities. *Philosophical Transactions of the Royal Society B, 364*(1526), 1–10. https://doi.org/10.1098/rstb.2008.0311

Ibrahim, K. I. M. (2017). The effect of using waste glass [WG] as partial replacement of sand on concrete. *ISOR Journal of Mechanical and Civil Engineering, 14*(2), 41–45. http://www.iosrjournals.org/iosr-jmce/papers/vol14-issue2/Version-2/G1402024145.pdf

Ismail, Z. Z., & Al-Hashmi, E. A. (2009). Recycling of waste glass as a partial replacement for fine aggregate in concrete. *Waste Management, 29*(2), 655–659. https://doi.org/10.1016/j.wasman.2008.08.012

Islam, G. M. S., Rahman, M. H., & Kazi, N. (2017). Waste glass powder as partial replacement of cement for sustainable concrete practice. *International Journal of Sustainable Built Environment, 6*(1), 37–44. https://doi.org/10.1016/j.ijsbe.2016.10.005

Ismail, S., Hoe. K. W., & Ramli, M. (2013). Sustainable aggregates: The potential and challenge for natural resources conservation. *Procedia—Social and Behavioral Sciences, 101*(1), 100–109. https://doi.org/10.1016/j.sbspro.2013.07.183

Jani, Y., & Hogland, W. (2014). Waste glass in the production of cement and concrete—A review. *Journal of Environmental Chemical Engineering, 2*(3), 1767–1775. https://doi.org/10.1016/j.jece.2014.03.016

Johari, A., Alkali, H., Hashim, H., Ahmed, S. I., & Mat, R. (2014). Municipal solid waste management and potential revenue from recycling in Malaysia. *Modern Applied Science, 8*(4), 37–49. https://doi.org/10.5539/mas.v8n4p37

Kadir, A. A., Ismail, M. R., Bakri, M. M. A., Sarani, N. A., & Hassan, M. I. H. (2016). The usage of glass waste as cement replacement. *Key Engineering Materials, 673*(1), 95–104. https://doi.org/10.4028/www.scientific.net/KEM.673.95

Khalil, W., & Al-Obeidy, N. (2018). Some properties of sustainable concrete containing two environmental wastes. *MATEC Web of Conference, 162*(1), 1–9. https://doi.org/10.1051/matecconf/201816202029

Khan, S. U., Nuruddin, M. F., Ayub, T., & Shafiq, N. (2014). Effects of different mineral admixtures on the properties of fresh concrete. *The Scientific World Journal, 1*, 1–11. https://doi.org/10.1155/2014/986567

Kou, S. C., & Xing, F. (2012). The effect of recycled glass powder and reject fly ash on the mechanical properties of fibre-reinforced ultrahigh performance concrete. *Advances in Materials Science and Engineering, 1*, 1–8. https://doi.org/10.1155/2012/263243

Kumar, M. H., Mohanta, N. R., Samataray, S., & Kumar, N. M. (2021). Combined effect of waste glass powder and recycled steel fibers on mechanical behavior of concrete. *SN Applied Science, 3*(1), 350. https://doi.org/10.1007/s42452-021-04353-6

Makul, N., Fediuk, R., Amran, M., Zeyad, A. M., Klyuev, S., Chulkova, I., Ozbakkaloglu, T., Vatin, N., Karelina, M., & Azevedo, A. (2021). Design strategy for recycled aggregate concrete: A review of status and future perspectives. *Crystals, 11*(6), 965. https://doi.org/10.3390/cryst11060695

Malek, M., Lasica, W., Jackowski, M., & Kandela, M. (2020). Effect of waste glass addition as a replacement for fine aggregate on properties of mortar. *Materials, 13*(14), 3189. https://doi.org/10.3390/ma13143189

Manomi, N., Sathyan, D., & Anand, K. B. (2018). Coupled effect of superplasticizer dosage and fly ash content on strength and durability of concrete. *Materials Today: Proceedings, 5*(11), 24033–24042. https://doi.org/10.1016/j.matpr.2018.10.196

Matschei, T., Lothenbach, B., & Glasser, F. P. (2007). The role of calcium carbonate in cement hydration. *Cement and Concrete Research, 37*(4), 551–448. https://doi.org/10.1016/j.cemconres.2006.10.013

Meddah, M. S. (2019). Use of waste window glass as substitute of natural sand in concrete production. *IOP Conference Series: Materials Science and Engineering, 603*(3), 1–9. https://doi.org/10.1088/1757-899X/603/3/032011

Mavroulidou, M. (2017). Mechanical properties and durability of concrete with water cooled copper slag aggregate. *Waste and Biomass Valorization, 8*(1), 1841–1854. https://doi.org/10.1007/s12649-016-9819-3

Mirzahosseini, M., & Riding, K. A. (2014). Effect of curing temperature and glass type on the pozzolanic reactivity of glass powder. *Cement and Concrete Research, 58*(1), 103–111. https://doi.org/10.1016/j.cemconres.2014.01.015

Mohammed, M. S., Mohamed, S. A., & Johari, M. A. M. (2016). Influence of superplasticizer compatibility on the setting time, strength and stiffening characteristics of concrete. *Advances in Applied Sciences, 1*(2), 30–36. https://doi.org/10.11648/j.aas.20160102.12

Needhidasan, S., Samuel, M., & Chidambaram, R. (2014). Electronic waste—An emerging threat to the environment of urban India. *Journal of Environmental Health Science and Engineering, 12*(1), 36. https://doi.org/10.1186/2052-336X-12-36

Olofinnade, O. M., Ede, A. N., Ndambuki, J. M., Egene, B. U., Akinwumi, I. I., & Ofuyatan, O. (2018). Strength and microstructure of eco-concrete produced using waste glass as partial and complete replacement for sand. *Coagent Engineering, 5*(1), 1483860. https://doi.org/10.1080/23311916.2018.1483860

Pacewska, B., & Wilinska, I. (2020). Usage of supplementary cementitious materials: Advantages and limitations. *Journal of Thermal Analysis and Calorimetry, 142*(1), 371–393. https://doi.org/10.1007/s10973-020-09907-1

Rashad, A. M. (2014). Recycled waste glass as fine aggregate replacement in cementitious materials based on Portland cement. *Construction and Building Materials, 72*(1), 340–357. https://doi.org/10.1016/j.conbuildmat.2014.08.092

Rasmussen, C., Vigso, D., Ackerman, F., Porter, R., Pearce, D., Dijkgraaf, E., & Vollebergh, H. (2005). *Rethinking the waste hierarchy* (pp. 1–118). Environmental Assessment Institute. https://www.osti.gov/etdeweb/servlets/purl/20623175

Safiuddin, M., Amrul Kaish, A. B. M., Woon, C.-O., & Raman, S. N. (2018). Early-age cracking in concrete: Causes, consequences, remedial measures, and recommendations. *Applied Sciences, 8*(10), 1730. https://doi.org/10.3390/app8101730

Sales, R. B. C., Sales, F. A., Figueiredo, E. P., Santos, W. J., Mohallem, N. D. S., & Aguilar, M. T. P. (2017). Durability of mortar made with fine glass powdered particles. *Advanves in Materials Science and Engineering, 1*, 1–9. https://doi.org/10.1155/2017/3143642

Sathyan, D., Anand, K. B., Mini, K. M., & Aparna, S. (2017). Optimization of superplasticizer in Portland pozzolana cement mortar and concrete. *IOP Conference Series: Materials Science and Engineering, 310*(1), 1–10. https://doi.org/10.1088/1757-899X/310/1/012036

Saraswathy, V., Karthick, S., Lee, H. S., Kwon, S.-J., & Yang, H.-M. (2017). Comparative study of strength and corrosion resistant properties of plain and blended cement concrete types. *Advances in Materials Science and Engineering, 1*, 1–14. https://doi.org/10.1155/2017/9454982

Singh, S., Srivastava, V., & Agarwal, V. C. (2015). Glass waste in concrete: Effect on workability and compressive strength. *International Journal of Innovative Research in Science, Engineering and Technology, 4*(9), 8142–8150.

Tariq, S., Scott, A. N., Mackechnie, J. R., & Shah, V. (2020). Durability of high volume glass powder self-compacting concrete. *Applied Sciences, 10*(22), 8058. https://doi.org/10.3390/app10228058

Tayeh, B. A., Shahidan, S., El-Kurd, J. A., Mohd Zuki, M., Ridzuan, A. R. M., Wan Ibrahim, M. H., Khalid, F. S., & Nazri, F. M. (2018). Utilization of waste glass powder as a partial replacement of cement in sustainable mortar practice. *Malaysian Construction Research Journal, 4*(2), 83–92.

Thomas, M. (2007). *Optimizing the use of fly ash in concrete* (pp. 1–24). Concrete. https://www.cement.org/docs/default-source/fc_concrete_technology/is548-optimizing-the-use-of-fly-ash-concrete.pdf

Wang, Y., Cao, Y., Zhang, P., & Ma, Y. (2020). Effective utilization of waste glass as cementitious powder and construction sand in mortar. *Materials, 13*(3), 707. https://doi.org/10.3390/ma13030707

Xun, W., Wu, C., Leng, X., Li, J., Xin, D., & Li, Y. (2020). Effect of functional superplasticizers on concrete strength and pore structure. *Applied Sciences, 10*(10), 3496. https://doi.org/10.3390/app10103496

Yaacob, I. I., Ali, M. Y., Sopyan, L., & Hashmi, S. (2015). Effect of calcium carbonate replacement on workability and mechanical strength of Portland cement concrete. *Advanced Materials Research, 1115*(1), 137–131. https://doi.org/10.4028/www.scientific.net/AMR.1115.137

Yu, P., Kirkpatrick, J., Poe, B., McMillan, P. F., & Cong, X. (2004). Structure of calcium silicate hydrate (C-S-H): Near-, mid-, and far-infrared spectroscopy. *Journal of the American Ceramic Society, 82*(3), 742–748. https://doi.org/10.1111/j.1151-2916.1999.tb01826.x

Zeng, X., Lan, X., Zhu, H., Liu, H., Umar, H. A., Xie, Y., Long, G., & Ma, C. (2020). A review on bubble stability in fresh concrete: Mechanisms and main factors. *Materials, 13*(8), 1820. https://doi.org/10.3390/ma13081820

Uncrushed Cockleshell as Coarse Aggregate Filler Replacement in Concrete

Chin Mei Yun, Md Rezaur Rahman, Kuok King Kuok, Amelia Chai Pei Sze, Joel Tiong Kung-Jiek, and Muhammad Khusairy Bin Bakri

Abstract This research aims to look at the effects of using uncrushed cockle shells (UCS) as a coarse aggregate in standard weight concrete (NWC). The local fish called it "blood cockle", is also known as "Kerang" in Malay. Landfill management issues and air pollution resulted from the disposal of large amounts of waste shells into landfills. As more and more mining activities for building aggregate are carried out, it was critical to investigate new possible materials that could turn waste into useable construction material. Hence, the idea of using waste cockle shells as one possible material is developed. In this study, the coarse aggregate was substituted with 10, 20, 30, 40, and 50% Uncrushed Cockle Shell (UCS). A total of 36 concrete cube specimens were produced in six batches with varying percentages of uncrushed cockle shell replacement. Six concrete cube samples per batch were cast and water cured for 7 and 28 days, respectively, before the compressive strength test. The workability and compressive strength between the Ordinary Portland Cement (OPC) and (UCS) concrete are compared in this study article. A slump cone test was performed on the (OPC) and (UCS) concrete mixtures to observe and record the slump value. Both OPC and (UCS) concrete cube specimens' compressive strength were evaluated. In addition, the best % replacement of cockle shells as coarse aggregate in terms of maximal compressive strength was found. Repurposing aquaculture by-products to replace natural aggregates would be a brilliant concept. However, putting this theory into practice is difficult since more supporting data are needed.

Keywords Aggregate · Cockle shell · Concrete · Workability · Compressive

C. M. Yun · K. K. Kuok · A. C. P. Sze · J. T. Kung-Jiek
Faculty of Engineering, Computing and Science, Swinburne University of Technology Sarawak Campus, Jalan Simpang Tiga, 93400 Kuching, Sarawak, Malaysia

M. R. Rahman (✉) · M. K. B. Bakri
Faculty of Engineering, Universiti Malaysia Sarawak, Jalan Datuk Mohammad Musa, 94300 Kota Samarahan, Sarawak, Malaysia
e-mail: rmrezaur@unimas.my

M. R. Rahman et al. (eds.), *Waste Materials in Advanced Sustainable Concrete*, Engineering Materials, https://doi.org/10.1007/978-3-030-98812-8_4

1 Introduction

Aggregate, ordinary Portland Cement (OPC), and water make up most of the content for concrete (Arimanwa et al., 2016; Collivignarelli et al., 2020; Uddin et al., 2013; Wegian, 2010). Concrete is a necessary building material used every day in our life due to its being strong, durable, and easy to precast into any shape or form (Asamoah et al., 2016). Concrete output in the world surpasses 10 billion tons per year (Eziefula, 2017). Concrete has become important building material for the construction industry has developed dramatically over the years (Benghida, 2017; Imbabi et al., 2012; Pheng & Hou, 2019; Wu et al., 2019). As a result, a high rate of aggregate demand leads to a high frequency of mining and quarrying operations (Csikosova et al., 2020; Langer & Arbogast, 2002; Rio-Chanona et al., 2020; Verschuur et al., 2021). Mining and quarrying operations would be a hazard to both natural habitat and natural minerals to maintain the demand rate for aggregate for concrete production (Drew et al., 2002; Langer & Arbogast, 2002; Poulin & Sinding, 1996). If there is no impact of control measures mining activities, the natural aggregate would be exhausted in the following few years (Qureshi et al., 2015).

The blood cockle with the scientific name *Anadara granosa*, also known as "Kerang" in Malay, is a popular seafood product in Malaysia's seafood sector (Hossen et al., 2014). At the local market, most of these *mollusk* species, i.e., oyster, scallop, mussel, and cockle, are common shellfish available at competitive prices (Hamli et al., 2019). Blood cockle is abundant in vitamin A, zinc, and protein, which may explain its popularity (Muthusamy & Sabri, 2012). Cockles were found along the nearby coastline and riverside in 2015, weighing up to 16,000 t (Ramakrishna & Sateesh, 2016). Once the cockle has been devoured at seafood or another restaurant, the by-product is disposed of in a landfill, resulting in an unpleasant odor, waste management issues, and expensive transportation and disposal costs (Alfagi et al., 2015). As a result of the negative consequences of the aquaculture by-product, efforts to use discarded shells into possible construction material are being made (Martinez-Porchas & Martinez-Cordova, 2012; Morris et al., 2018; Sadh et al., 2018).

Few similar studies have utilized cockleshells to partially replace coarse and fine material (Eziefula, 2017). *Mollusc* output in 2014 was 16,113 thousand tons, accounting for 21.8% of global aquaculture production. The fine or coarse pebbles used to replace 25% of the seashells often exceed 20 N/mm^2. According to this study, fine aggregate containing 10–20% crushed cockle shell had a maximum compressive strength value that surpassed the regulated concrete strength on the 28th day. The surface roughness of the cockleshell is rougher than granite aggregate, which improves bonding and increases inter-particle friction, resulting in increased concrete compressive strength (Eziefula, 2017; Muthusamy & Sabri, 2012; Ponnada et al., 2016; Saranya, 2017).

The debonding of cement paste and aggregate particles caused the failure of normal weight concrete (Barham et al., 2019; Hilal et al., 2016; Rico et al., 2017; Shafigh et al., 2012). Because flaky, elongated, angular, and abrasive particles contain more void spaces, the combination needs more water to improve workability and additional sand to fill the blank spaces (Polat et al., 2013). Flaky and elongated particles create abrasive mixes that are difficult to deal itself. Micro-cracks propagate in the mortar-aggregate interface with lower strength concrete. In contrast, micro-cracks propagate through the aggregate in higher strength concrete because the bonding strength between the mortar-aggregate interface is larger (Qureshi et al., 2015). As a result, the interfacial bond strength of the mortar-aggregate contact varies since various aggregate types interact differently with the mortar. As a result, depending on the aggregate type and the aggregate-mortar bonding strength, microcracks might propagate at the mortar joint or through the aggregate itself (Czapik, 2020; Rashad, 2016; Zhang et al., 2013).

A crushed cockle shell with a particle size of 10–14 mm was used as coarse aggregate in concrete throughout the study made by Muthusamy and Sabri (2012). In the experiment, the water-to-cement ratio was kept constant at 0.5. The mix with 20% crushed cockle shell replacement had the maximum compressive strength of approximately 35 MPa, higher than other mixes, including the control concrete, which had a compressive strength of 30 MPa (Muthusamy & Sabri, 2012). Replacement of fine aggregate with crushed cockle shell at 10% replacement yielded the highest compressive strength (Ramakrishna & Sateesh, 2016). The crushed cockle shell was used as a space filler, which increased strength development. Crushed cockle shell as fine aggregate and uncrushed cockle shell as coarse aggregate at a 25% replacement level resulted in compressive strength of 49.65 MPa, which was higher than the control concrete's compressive strength 48.22 MPa (Ramakrishna & Sateesh, 2016).

Partial substitution of coarse aggregate in the particle size ranges of 10-20 mm resulted in a reduction in concrete compressive strength (Ponnada et al., 2016). It might be owing to the cockle shell's high crushing and impact value. A batch with 15% crushed cockleshell replacement in the coarse aggregate and 20% granite powder replacement in the fine aggregate had a compressive strength of 43.7 MPa, greater than the conventional concrete's 30 MPa. IT might be owing to the high silica content of the granite powder, which created a layer on top of the crushed shell, resulting in the creation of C–S–H gel (Ponnada et al., 2016). The *Scapharca subcrenata* shell's radial rib nodules are comparable to the cockleshell in that the complex structure may absorb energy at the impact of external forces (George et al., 2019). Furthermore, the radial ribs with nodules on the shell's surface can lower impact velocity and abrasive energy particles, resulting in less micro-crack propagation. This complex structure may reduce microcrack propagation, fatigue wear, and fracture micro-mechanical wear (Tian et al., 2014).

There have been no similar experimental investigations done in the past with this research study. Crushed or uncrushed cockle shells of various shapes can be used to substitute fine or coarse material. This research study used an uncrushed cockle shell (UCS) with a 20–40 mm particle size to partially replace coarse aggregate. The results of the workability and compressive strength of concrete containing (UCS) are still unknown. As a result, this study aims to find the ideal water-cement ratio for workability and determine the highest compressive strength value with the best percentage replacement. The focus of this study is for partial replacement of coarse aggregate with uncrushed cockle shells for varying percentages. This study aims to establish the maximum compressive strength of (UCS) concrete with the best percentage of replacement and evaluate the workability of (UCS) concrete with a given water-cement ratio. In addition, the study's goal was to raise public awareness about the notion of repurposing aquaculture waste as a viable construction material. This idea would not only solve the disposal problem but would also minimize natural aggregate quarrying.

2 Methodology

2.1 *Materials*

Concrete materials such as Ordinary Portland cement were obtained from Cahya Mata Sarawak (CMS) Berhad. The uncrushed cockle shell waste was obtained from a local seafood restaurant. River sand as fine aggregates and gravel as coarse aggregates were obtained from a local supplier.

2.2 *Sample Preparation*

Before the casting, the cast ingredients were carefully prepared. Cockle shells were gathered from seafood restaurants and cleaned, steeped in vinegar, and dried. As indicated in Fig. 1, the Uncrushed Cockle Shells (UCS) were sieved into 20–40 mm particle size and weighted before casting. The fine aggregate adopted was river sand, while the coarse aggregate used was gravel. The gravel and uncrushed cockle shells (UCS) were sieved into various size ranges of 20, 25, 28, 34, and 37.5 mm. The mixer machine was used to mix concrete ingredients such as Portland Cement, uncrushed cockle shell, fine and coarse aggregates. The water was added, and the mixing took three minutes. Following the mixing, fresh concrete was poured into concrete molds. The concrete cubes were cast and then de-molded for a water cure after 24 h. The compressive strength of the cast concrete cube specimens was evaluated after being water cured for 7 and 28 days. Binder was made of Portland cement. Figure 2 shows the sample preparation flowchart.

2.3 *Nominal Proportion*

By volume, the nominal proportions were 1:1.62:4.99 (cement: fine aggregate: coarse aggregate). This experiment has six batches, each with a different percentage of Uncrushed Cockle Shell (UCS) substitution for coarse aggregate: 0, 10, 20, 30, 40, and 50%. Fresh concrete was compressed with a tamping rod to reduce segregation

Fig. 1 Uncrushed cockle shell as coarse aggregate

and air bubble entrapment during the casting process. The compressive strength of the hardened concrete cubes was measured after 7 and 28 days of water curing. The findings of the experiments were documented and discussed in the following section. Table 1 summarizes the results of the experimental study.

2.4 Slump Test

A slump test was performed for each sample before casting to assess the workability of each kind of combination using a slump cone, base plate, tamping rod, and ruler. As a guide, the ASTM C143/C143M-20 (2020) standard was employed.

2.5 Compressive Strength Test of Concrete

After 7 and 28 days of curing, each cube was tested for compressive strength. For each batch, the average results were obtained from the three cubes cast. The compressive strength test was performed using Unit Test Autocon 2000 in compliance with BS EN 12,390-3 (2019). During the compressive strength test, the load was delivered

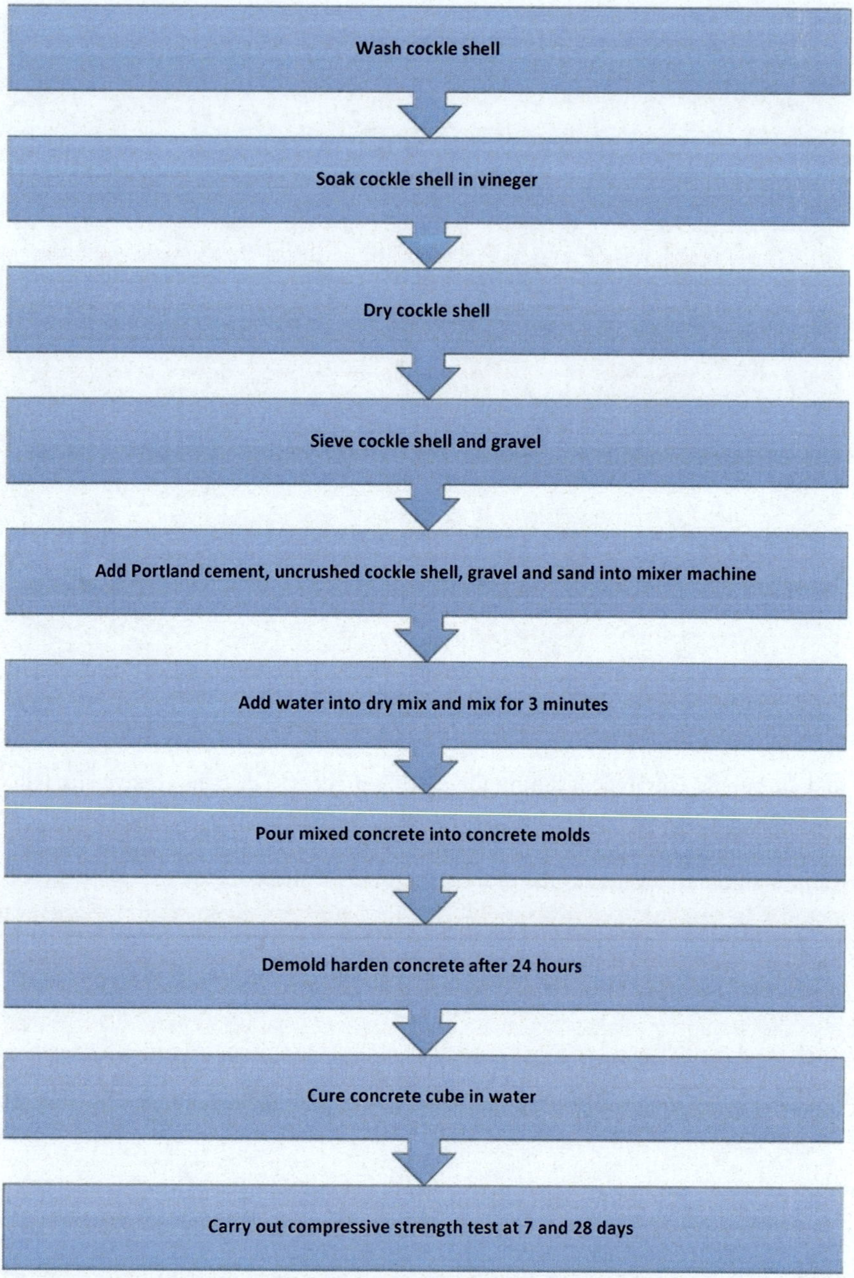

Fig. 2 Procedures for the uncrushed cockle shell concrete sample preparation

Table 1 Concrete sample

Batch	Percentage of uncrushed cockle shell (%)	Percentage of coarse aggregate (%)	Compressive strength test	
			7 days	28 days
1 OPCC C0	0	100	3	3
2 UCSC C1	10	90	3	3
3 UCSC C2	20	80	3	3
4 UCSC C3	30	70	3	3
5 UCSC C4	40	60	3	3
6 UCSC C5	50	50	3	3

to the mortar specimens gradually at 0.05 ± 0.01 kN/s across an area of 100 mm × 100 mm until the specimens failed. The results were averaged and reported from mortar specimens from the same batch.

2.6 *Scanning Electron Microscope (SEM) and Energy Dispersive X-Ray/Spectroscopy (EDS/EDX)*

SEM and EDX/EDS were carried out in compliance with ASTM C1723-16 (2016) and ASTM E1508-12 (2019) standards, respectively. 100×, 250×, 500×, and 10,000× magnifications were used to analyze the samples. Automated software scans and analyses the elemental composition percentages of the samples. The EDX/EDS was conducted on each sample many times at various phases, with the most representative results chosen. Hitachi TM4000Plus Tabletop Microscope (Hitachi Ltd., Tokyo, Japan) with a Quantax75TM Series Energy Dispersive X-Ray Spectrometer was used in the SEM and EDX/EDS investigations of the materials.

2.7 *Fourier Transform Infrared (FTIR) Spectroscopy*

Fourier-transform infrared spectroscopy was used to examine the materials (IRAffinity-1, Shimadzu Corporation, Kyoto, Japan). Fourier-transform infrared spectroscopy was used for qualitative and quantitative analysis by ASTM E168-16 (2016) standards. The spectra of each sample were scanned in the range of 4000–400 cm^{-1}. The sample's infrared spectrum transmittance and absorption were utilized to generate a unique chemical fingerprint spectrum using Fourier-transform infrared spectroscopy. The test was repeated multiple times for each sample, and the most representative results were chosen.

3 Results and Discussions

3.1 Workability Properties

The workability of fresh concrete was assessed using the slump cone test, with the water-cement ratio set at 0.55. The slump's form was noted, as well as the slump's value. The result for new concrete workability is shown in Fig. 3.

Figure 3 demonstrates that the slump value of fresh concrete increases as the percentage of (UCS) used to replace coarse aggregate increases. When compared to the controlled batch (OPCC C0) with 141 mm slump height, batch (C5) with 50% (UCS) replacement had the greatest slump value of 175 mm. Two primary elements impact the workability of fresh concrete in this experiment. In this experiment, the coarse aggregate surface roughness was the first element that influenced workability. The cockleshell has a smooth inner surface and a rough outer surface. The smooth and flat inner surface of the cockleshell and mortar created low frictional force and required less mortar to cover the surface area of the cockleshell. It contributes to the higher workability of cockleshell concrete. According to Adullahi (2012), smooth and round aggregates improved the workability of fresh concrete by requiring less cement paste to cover the aggregate surface. The controlled batch (C0) and cockle

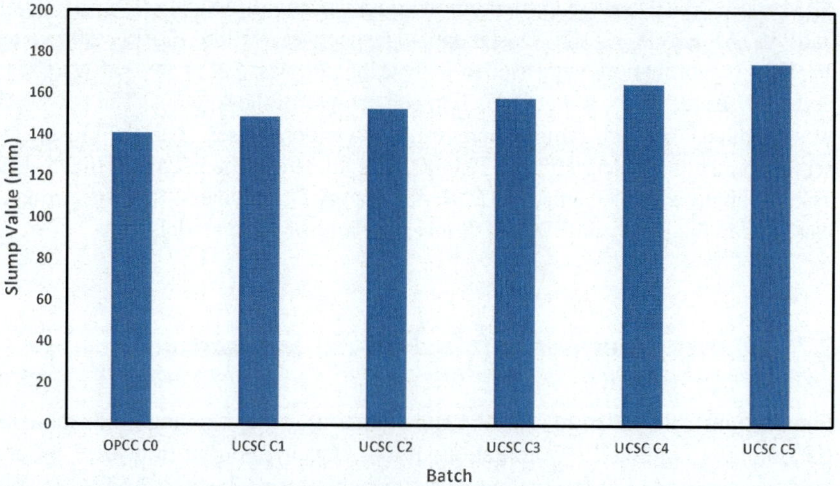

Fig. 3 Slump with a different batch of concrete specimens

Fig. 4 Slump cone test **a** OPCC, and **b** UCSC C5

shell concrete are shown in Figs. 4a, b. The controlled batch, which utilized 100% gravel coarse aggregate and had a lower slump value than the other mix batches, as shown in Fig. 4a, had a lower slump value than the other mix batches. Gravel has a rougher surface, which causes a more frictional force to be created during compaction in the slump cone test. As a result, gravel adheres better to the mortar, resulting in more compacted new concrete and a lower slump value.

On the other hand, the aggregate form is the second component that affects the workability of new concrete. As spherical aggregate has a smaller surface area and frictional force, it has a greater slump value than flat and elongated aggregate, according to Polat et al. (2013). In addition, angular aggregate has a greater water requirement than spherical aggregate because of its uneven form and abrasive surface roughness. Because the shell of a univalve cockle is semi-spherical in form, less mortar is required to cover it because the shell's surface area is tiny. Because less mortar is required to coat the shell, the mortar in fresh concrete produced a greater lubricating effect, as shown in Fig. 4b. Because less frictional force is created during compaction in the slump cone test, this phenomenon improved the workability of (UCS) new concrete. The rough surface of the shell does not harm fresh concrete, while the smooth surface of the shell improves workability. Fresh concrete semi-spherical cockle shell form offers greater workability than fresh concrete with flat, elongated, and irregular cockle shell shape.

3.2 Compressive Strength Properties

In general, as the cockleshell partially replaced coarse aggregate, the compressive strength of concrete decreased, as shown in Fig. 5. At 7 days strength, control concrete (OPC C0) has the highest compressive strength of 21.82 MPa. Among UCS concrete, UCS C3 with 30% cockle shell obtained the highest compressive strength of 20 MPa, while USC C4 with 40% cockle shell obtained the lowest compressive strength of 18.34 MPa. Concrete failure occurs in the interfacial transition zone (ITZ) during the early stage. The cement hydration was not fully activated. Hence microcracks spread along with the aggregate-matrix contact when pressure is applied.

At 28 days, the compressive strength increased as the cockleshell replacement percentage increased from 10 to 30%. As a cockle shell replacement percentage further increased from 30 to 50%, the compressive strength decreased. UCS C3 with a 30% cockle shell obtained the highest compressive strength of 33.82 MPa, followed by control concrete (OPC C0) with compressive strength of 31.45 MPa. The lowest compressive strength was 25.57 MPa for USC C5 with a 50% cockle shell. The optimal replacement percentage of cockleshells was 30% and was able to outperform control concrete. The failure mode of UCS concrete was observed to be cracking through the cockleshell and gravel, as shown in Fig. 6.

In this experiment, the replacement of coarse aggregate was based on weight. Since gravel was denser than cockle shell, cockleshell has a higher volume than gravel at the same weight, thus having a greater specific surface area. As a result,

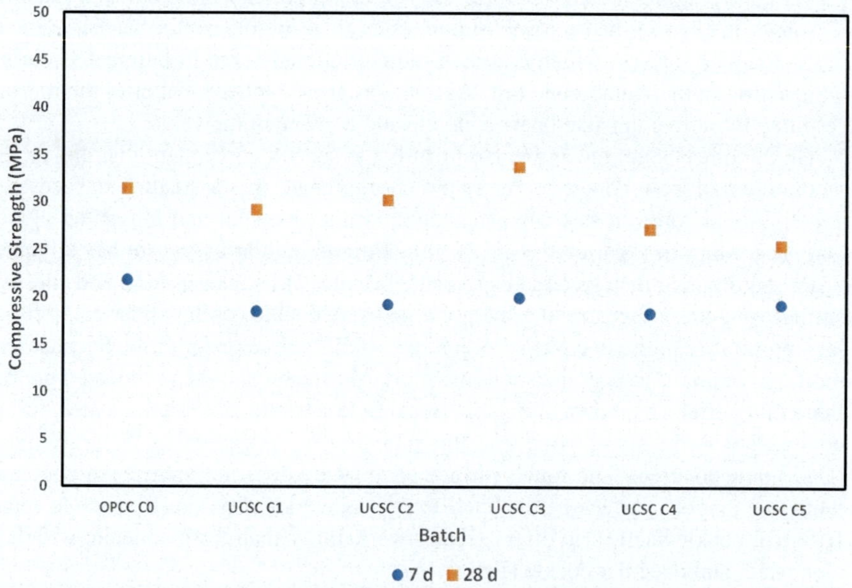

Fig. 5 Compressive strength with a different batch of concrete specimens

Fig. 6 Failure mode for 30% coarse cockle aggregate (UCS C3) concrete of 28-days data

additional cement paste was required to coat and bond to the cockleshell. The water-to-cement ratio was set at 0.55. To ensure the workability of the concrete mix. It is the resulting dilution of cement paste that reduces the bonding strength of mortar and aggregate.

3.3 Morphological Properties

Figure 7a–d show SEM images of the UCS C3 (28 days) at 100×, 250×, 500×, and 1000× magnifications, respectively. The flaky strong topology of the surface sample is seen in Fig. 7. As it mixed within themselves or with other materials, the hydraulic reactivity, or chemical or physical contact, was minimal, making it difficult to produce strong interlock with small particles (Rahman et al., 2017). The appropriate paste combination, soundness, and hydraulic reactivity were responsible for the compact microstructure (Chin et al., 2020a, 2020b; Olivia et al., 2017). After a longer curing period, the flaky microstructure became highly thick and coherent, indicating that the cementitious bond virtually completely covered the sample with fewer pore spaces (Chin et al., 2020a, 2020b; Olivia et al., 2016).

The problem with using coarse cockleshells is that one side surface of the shell is smooth, and another side of the surface is rough (Mohamad et al., 2021). The smooth side of the cockleshell usually created less bonding with paste, while the

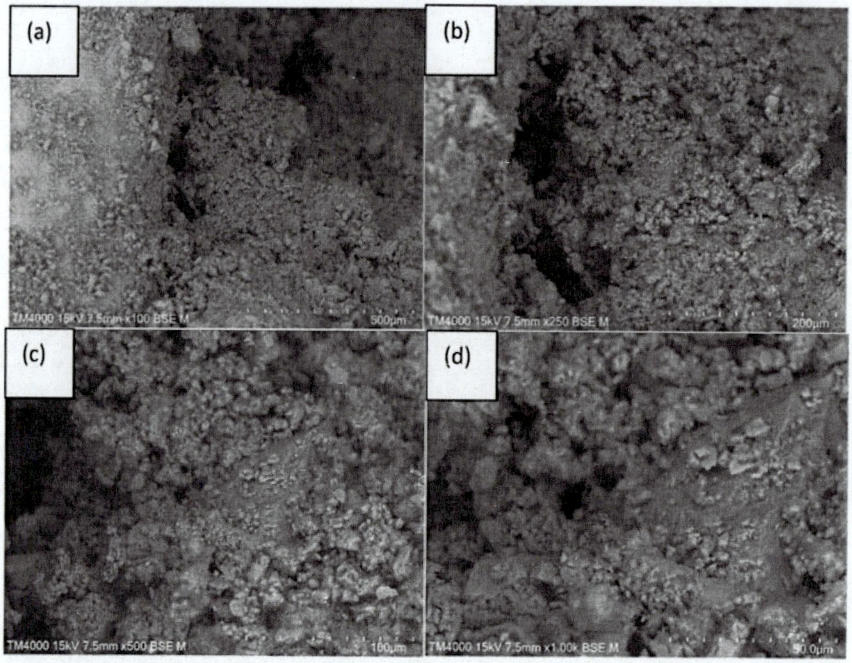

Fig. 7 SEM images for UCS C3 (28 days) at magnification of **a** 100×, **b** 250×, **c** 500×, and **d** 1000×

rough side usually bonded strongly with the paste. The smooth part initially began to slip and became the initial crack for the concrete, which caused the primary failure in concrete. In addition, the density differences caused most of the cockleshells to float and sink in the upper and bottom parts of the concrete. This improper distribution cause problem in distributing stress on the concrete.

3.4 *Elemental Properties*

Table 2 and Fig. 8 show the EDS/EDX plot and element contents for UCS C3 (28 days). The elements present in the sample, according to Table 2, were carbon (C), oxygen (O), silicon (Si), aluminum (Al), and calcium (Ca). Table 2 shows that calcium mass percentage has the highest content, followed by calcium and carbon. It shows that chemical reactions happen during the mixed of paste mixture and hydraulic reactivity, which create calcium carbonate ($CaCO_3$) and the presence of cockle element (Matschei et al., 2007; Yaacob et al., 2015). $CaCO_3$ helps to improve early strength by having an accelerator effect and a high rate of hydration, which causes the concrete to solidify faster. When concrete with $CaCO_3$ has developed, it has a lesser strength than concrete without $CaCO_3$, although it is still within the

Table 2 Element composition of UCS C3 (28 days)

Element	Mass normal (%)	Atom (%)
Ca	48.76	27.34
O	39.49	55.46
C	7.23	13.53
Si	3.15	2.52
Al	1.36	1.13
Sum	100.00	100.00

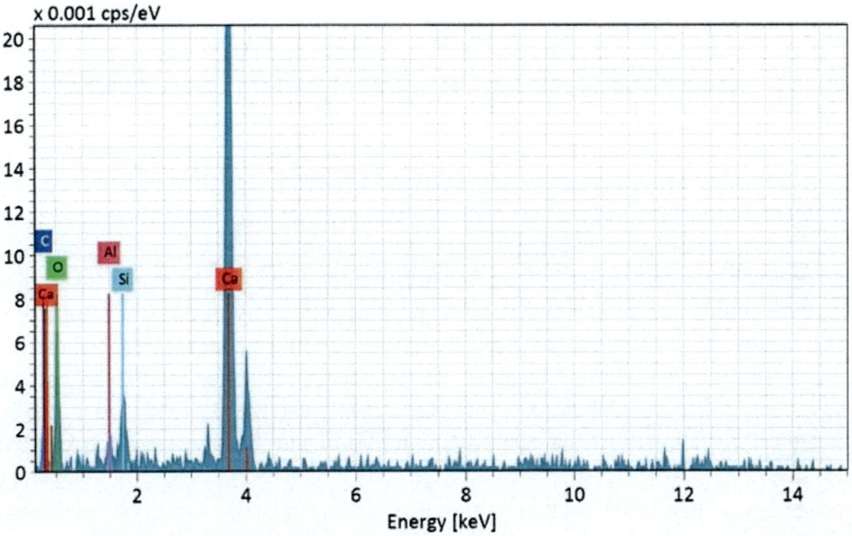

Fig. 8 EDS/EDX graph for UCS C3 (28 days)

required strength range (Matschei et al., 2007; Yaacob et al., 2015) as evidenced by SEM morphological pictures, appropriate mixing ratio, and chemical substance used to generate good reactivity, which creates a strong connection between each material employed in the combination for the sample.

3.5 *Infrared Spectral Properties*

Figure 9 depicts the FTIR of UCS C3 (28 days). The O–H bond, which was owing to hydraulic of water bond and moisture, produced three small band peaks at 3200–3600 cm^{-1}. The peak at 2549.89, 2185.97, and 2005.97 cm^{-1} was attributable to C-H bending and stretching (Islam et al., 2011). The weak bands at 1716.55 and 1496.76 cm^{-1} corresponded to C=O bonds from carbonate (Kang et al., 2008;

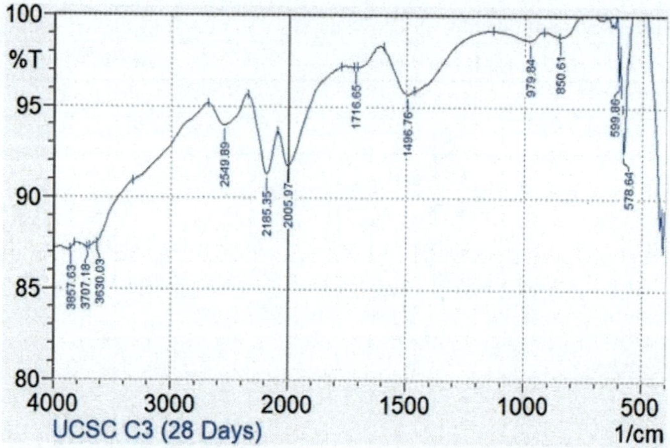

Fig. 9 FTIR spectra of UCS C3 (28 days)

Witoon, 2011). Due to symmetric and asymmetric valence and deformation vibration, the most frequent wavelengths are 979.84, 850.61, 599.86, and 578.64 cm^{-1}, respectively (Kamba et al., 2013; Vaiciukyniene et al., 2013). Some researchers mentioned that calcium carbonate could be well defined by infrared at band 979.84 and 850.61 cm^{-1} (Doostmohammadi et al., 2011). While some researchers noted that the presence of aragonite in the cockleshell was also confirmed at 979.84 cm^{-1} of $CO_3{}^{2-}$ symmetric stretching (Gergely et al., 2010; Islam et al., 2011; Zakaria et al., 2008). At the same time, the sharp band at 850.61 cm^{-1} is related to Ca-O bonds (Gergely et al., 2010).

4 Conclusion

In conclusion, this research contributes to a better knowledge of the impact of utilizing uncrushed cockleshells (UCS) as a coarse aggregate replacement in concrete. As the percentage replacement of (UCS) increases, the workability of the concrete mixture improves. The optimal percentage replacement of (UCS) in concrete is 30%, with a maximum compressive strength of 33.82 MPa, 7.53% greater than the maximum compressive strength of (OPC) Concrete, which is 31.45 MPa. The ribs with nodules produced on the shell have shown to be resistant to micro-crack propagation, with (UCS) concrete showing better strength at 30% replacement. The compressive strength of the (UCS) concrete was comparable to that of the (OPC) concrete. The use of cockle shells as coarse aggregate may appear to be favourable. However, there are few barriers to using aquaculture by-products as concrete aggregates, such as accessibility, availability, transportation, and storage.

Acknowledgements The authors would like to thank Swinburne University of Technology Sarawak Campus and Universiti Malaysia Sarawak (UNIMAS) for the collaboration.

References

Abdullahi, M. (2012). Effect of aggregate type on compressive strength of concrete. *International Journal of Civil and Structural Engineering, 2*(3), 791–800.

Alfagi, A., Purnaweni, H., Setiani, O. (2015). Waste management in KFC and Pizza Hut case study in Semarang, Indonesia. *Science Journal of Environmental Engineering Research, 2015*(1), 1–15. https://www.sjpub.org/sjeer/sjeer-285.pdf

Arimanwa, M. C., Onwuka, D. O., & Arimanwa, J. I. (2016). Effect of chemical composition of ordinary Portland cement on the compressive strength of concrete. *International Refereed Journal of Engineering and Sciences, 5*(3), 20–31. http://www.irjes.com/Papers/vol5-issue3/D532031.pdf

ASTM E1252-98. (2013). *Standard practice for general techniques for obtaining infrared spectra for qualitative analysis.* ASTM International. https://doi.org/10.1520/E1252-98R13E01

ASTM E168-16. (2016). *Standard practices for general techniques of infrared quantitative analysis.* ASTM International. https://doi.org/10.1520/E0168-16

ASTM C1723-16. (2016). *Standard guide for examination of hardened concrete using scanning electron microscopy.* ASTM International. https://doi.org/10.1520/C1723-16

ASTM E1508-12. (2019). *Standard guide for quantitative analysis by energy-dispersive spectroscopy.* ASTM International. https://doi.org/10.1520/E1508-12AR19

ASTM C143/C143M-20. (2020). *Standard test method for slump of hydraulic-cement concrete.* ASTM International. https://doi.org/10.1520/C0143_C0143M-20

Asamoah, R. O., Ankrah, J. S., Offei-Nyako, K., & Tutu, E. O. (2016). Cost analysis of precast and cast-in-place concrete construction for selected public buildings in Ghana. *Journal of Construction Engineering.* https://doi.org/10.1155/2016/8785129

Barham, W. S., Obaidat, Y. T., & Maabreh, A. I. (2019). Effect of aggregate size on the bond behavior between carbon fiber-reinforced polymer sheets and concrete. *Journal of Materials in Civil Engineering, 13*(12), 1–10. https://doi.org/10.1061/(ASCE)MT.1943-5533.0002950

Benghida, D. (2017). Concrete as a sustainable construction material. *Key Engineering Materials, 744*(1), 196–200. https://doi.org/10.4028/www.scientific.net/KEM.744.196

BS EN 12390-3. (2019). *Testing hardened concrete. Compressive strength of test specimens.* British Standard Institution.

Chin, M. Y., Rahman, M. R., Yong, C. W. P., Wong, A. M. C., & Bakri, M. K. B. (2020a). The curing times effect on the strength of ground granulated blast furnace slag (GGBFS) mortar. *Construction and Building Materials, 260*(1), 120622. https://doi.org/10.1016/j.conbuildmat.2020.120622

Chin, M. Y., Rahman, M. R., Kuok, K. K., Chiew, W. Y., & Bakri, M. K. B. (2020b). Characterization and impact of curing duration on the compressive strength of coconut shell coarse aggregate in concrete. *BioResources, 16*(3), 6057–6073. https://ojs.cnr.ncsu.edu/index.php/BioRes/article/view/BioRes_16_3_6057_Chin_Curing_Duration_Compressive_Strength_Coconut_Shell

Collivignarelli, M., Cillari, G., Ricciardi, P., Miino, M. C., Toretta, V., Rada, E. C., & Abba, A. (2020). The production of sustainable concrete with the use of alternative aggregates: A review. *Sustainability, 12*(19), 7903. https://doi.org/10.3390/su12197903

Csikosova, A., Janoskova, M., & Culkova, K. (2020). Application of discriminant analysis for avoiding the risk of quarry operation failure. *Journal of Risk and Financial Management, 13*(10), 231. https://doi.org/10.3390/jrfm13100231

Czapik, P. (2020). Microstructure and degradation of mortar containing waste glass aggregate as evaluated by various microscopic techniques. *Materials, 13*(9), 2186. https://doi.org/10.3390/ma13092186

Doostmohammadi, A., Monshi, A., Salehi, R., Fathi, M. H., Golniya, Z., & Daniels, A. U. (2011). Bioactive glass nanoparticles with negative zeta potential. *Ceramics International, 37*(7), 2311–2316. https://doi.org/10.1016/j.ceramint.2011.03.026

Drew, L. J., Hanger, W. H., & Sachs, J. S. (2002). Environmentalism and natural aggregate mining. *Natural Resources Research, 11*(1), 19–28. https://doi.org/10.1023/A:1014283519471

Eziefula, U. G. (2017). Developments in utilisation of agricultural and aquaculture by-products as aggregate in concrete—A review. *Environmental Technology Reviews, 7*(1), 19–45. https://doi.org/10.1080/21622515.2017.1423399

Gergely, G., Weber, F., Lukacs, I., Toth, A. L., Horvath, Z. E., Mihaly, J., & Balazsi, C. (2010). Preparation and characterization of hydroxyapatite from Eggshell. *Ceramics International, 36*(2), 803–806. https://doi.org/10.1016/j.ceramint.2009.09.020

George, P., Abdul Hamid, Z., Bakar, M. Z. A., Perimal, E. K., & Bharatham, H. (2019). A short review on cockle shells as biomaterials in the context of bone scaffold fabrication. *Sains Malaysiana, 48*(7), 1539–1545. https://doi.org/10.17576/jsm-2019-4807-23

Hamli, H., Idris, M. H., Hena, M. K. A., & Rajaee, A. H. (2019). Fisheries assessment, gametogenesis and culture practice of local bivalve: A review. *Tropical Agricultural Science, 42*(1), 103–124. http://www.pertanika.upm.edu.my/resources/files/Pertanika%20PAPERS/JTAS%20Vol.%2042%20(1)%20Feb.%202019/08%20JTAS-1427-2017.pdf

Hilal, A. A., Thom, N. H., & Dawson, A. R. (2016). Failure mechanism of foamed concrete made with/without additives and lightweight aggregate. *Journal of Advanced Concrete Technology, 14*(9), 511–520. https://doi.org/10.3151/jact.14.511

Hossen, M. F., Hamdan, S., & Rahman, M. R. (2014). Cadmium and lead in Blood Cockle (Anadara granosa) from Asajaya, Sarawak, Malaysia. *The Scientific World Journal, 1*, 1–4. https://doi.org/10.1155/2014/924360

Imbabi, M. S., Carrigan, C., & McKenna, S. (2012). Trends and developments in green cement and concrete technology. *International Journal of Sustainable Built Environment, 1*(2), 194–216. https://doi.org/10.1016/j.ijsbe.2013.05.001

Islam, K. N., Bakar, M. Z. B. A., Noordin, M. M., Hussein, M. Z. B., Rahman, N. S. B. A., & Ali, M. E. (2011). Characterisation of calcium carbonate and its polymorphs from cockle shells (Anadara granosa). *Poweder Technologu, 213*(1–3), 188–191. https://doi.org/10.1016/j.powtec.2011.07.031

Kamba, A. S., Ismail, M., Ibrahim, T. A. T., & Zakaria, Z. A. (2013). Synthesis and characterisation of calcium carbonate aragonite nanocrystals from cockle shell powder (*Anadara granosa*). *Journal of Nanomaterials, 1*, 1–9. https://doi.org/10.1155/2013/398357

Kang, S.-P., Lee, J.-W., & Ryu, H.-J. (2008). Phase behaviour of methane and carbon dioxide hydrates in meso- and macro-sized porous media. *Fluid Phase Equilibria, 274*(1), 68–72. https://doi.org/10.1016/j.fluid.2008.09.003

Langer, W. H., & Arbogast, B. F. (2002). Environmental impacts of mining natural aggregate. In: A.G. Fabbri, G. Gaál, & R.B. McCammon (Eds.) *Deposit and geoenvironmental models for resource exploitation and environmental security* (Vol. 80, pp. 151–169). Nato Science Partnership Subseries: 2. Springer. https://doi.org/10.1007/978-94-010-0303-2_8

Matschei, T., Lothenbach, B., & Glasser, F. P. (2007). The role of calcium carbonate in cement hydration. *Cement and Concrete Research, 37*(4), 551–448. https://doi.org/10.1016/j.cemconres.2006.10.013

Martinez-Porchas, M., & Martinez-Cordova, L. R. (2012). World aquaculture: Environmental impacts and troubleshooting alternatives. *The Scientific World Journal, 1*, 1–9. https://doi.org/10.1100/2012/389623

Mohamad, N., Muthusamy, K., & Ismail, M. A. (2021). Cockle shell as mixing ingredient in concrete: A review. *Construction, 1*(2), 1–15. https://doi.org/10.15282/construction.v1i2.6503

Morris, J. P., Backeljau, T., & Chapelle, G. (2018). Shells from aquaculture: A valuable biomaterial, not a nuisance waste product. *Reviews in Aquaculture, 11*(1), 42–57. https://doi.org/10.1111/raq.12225

Muthusamy, K., & Sabri, N. A. (2012). *International Journal of Science, Environment and Technology, 1*(4), 260–267.

Olivia, M., Oktaviani, R., & Ismeddiyanto. (2017). Properties of concrete containing ground waste cockle and clam seashells. *Procedia Engineering, 171*(1), 658–663. https://doi.org/10.1016/j.proeng.2017.01.404

Pheng, L. S., & Hou, L. S. (2019). The economy and the construction industry. In: L. S. Pheng & L. S. Hou (Eds.), *Construction quality and the economy* (pp. 21–54). https://doi.org/10.1007/978-981-13-5847-0_2

Polat, R., Yadollahi, M. M., Sagsoz, A. E., & Arasan, S. (2013). The Correlation between Aggregate Shape and Compressive Strength of Concrete: Digital Image Processing Approach. *International Journal of Structural and Civil Engineering Research, 2*(3), 62–80. http://www.ijscer.com/uploadfile/2015/0429/20150429074945165.pdf

Ponnada, M. R., Prasad, S. S., & Dharmala, H. (2016). Compressive strength of concrete with partial replacement of aggregates with granite powder and cockle shell. *Malaysian Journal of Civil Engineering, 28*(2), 183–204.

Poulin, R., & Sinding, K. (1996). A North American perspective on land use and mineral aggregate production. *GeoJournal, 40*(3), 273–281. https://www.jstor.org/stable/41147000

Qureshi, M. A., Aslam, M., Shah, S. N. R., & Otho, S. H. (2015). Influence of aggregate characteristics on the compressive strength of normal weight concrete. *Technical Journal University of Engineering and Technology, 20*(3), 1–11.

Rahman, M. A., Sarker, P. K., Shaikh, F. U. A., & Saha, A. K. (2017). Soundness and compressive strength of Portland cement blended with ground granulated ferronickel slag. *Construction and Building Materials, 140*(1), 194–202. https://doi.org/10.1016/j.conbuildmat.2017.02.023

Ramakrishna, B., & Sateesh, A. (2016). Exploratory study on the use of cockle shell as partial Coarse & Fine aggregate replacement in concrete. *International Research Journal of Engineering and Technology, 3*(6), 2347–2349. https://www.irjet.net/archives/V3/i6/IRJET-V3I6440.pdf

Rashad, A. M. (2016). A comprehensive overview about recycling rubber as fine aggregate replacement in traditional cementitious materials. *International Journal of Sustainable Built Environment, 5*(1), 46–82. https://doi.org/10.1016/j.ijsbe.2015.11.003

Rico, S., Farshidpour, R., & Tehrani, F. M. (2017). State-of-the-art report on fiber-reinforced lightweight aggregate concrete masonry. *Advances in Civil Engineering, 1*, 1–9. https://doi.org/10.1155/2017/8078346

Rio-Chanona, R. M. D., Mealy, P., Pichler, A., Lafond, F., & Farmer, J. D. (2020). Supply and demand shocks in the COVID-19 pandemic: An industry and occupation perspective. *Oxford Review of Economic Policy, 36*(1), S94–S137. https://doi.org/10.1093/oxrep/graa033

Sadh, P. K., Duhan, S., & Duhan, J. S. (2018). Agro-industrial wastes and their utilization using solid state fermentation: A review. *Bioresources and Bioprocessing, 5*(1), 1–15. https://doi.org/10.1186/s40643-017-0187-z

Saranya, C. V. (2017). Assessment of concrete by partial replacement of coarse aggregate by cockle shell with glass fibre. *International Journal of Intelligence Research, 9*(1), 1–6. http://ijoir.com/Files/54.pdf

Shafigh, P., Jumaat, M. Z., & Mahmud, H. B. (2012). Effect of replacement of normal weight coarse aggregate with oil palm shell on properties of concrete. *Arabian Journal of Science and Engineering, 37*(1), 955–964. https://doi.org/10.1007/s13369-012-0233-2

Tian, L., Tian, X., Hu, G., Wang, Y., & Ren, L. (2014). Effects and mechanisms of surface topography on the antiwear properties of molluscan shells (*Scapharca subcrenata*) using the fluid-solid interaction method. *The Scientific World Journal, 1*, 1–12. https://doi.org/10.1155/2014/185370

Uddin, M. A., Jamee, M., Sobuz, H. R., Islam, M. S., & Hasan, N. M. S. (2013). Experimental study on strength gaining characteristics of concrete using Portland composite cement. *KSCE Journal of Civil Engineering, 17*(1), 789–796. https://doi.org/10.1007/s12205-013-0236-x

Vaiciukyniene, D., Skipkiunas, G., Dauksys, M., & Sasnauskas, V. (2013). Cement hydration with zeolite-based additive. *Chemija, 24*(4), 271–278.

Verschuur, J., Koks, E. E., & Hall, J. W. (2021). Global economic impacts of COVID-19 lockdown measures stand out in high-frequency shipping data. *PLoS ONE, 16*(4), e0248818. https://doi.org/10.1371/journal.pone.0248818

Wegian, F. M. (2010). Effect of seawater for mixing and curing on structural concrete. *The IES Journal Part a: Civil & Structural Engineering, 3*(4), 235–243. https://doi.org/10.1080/19373260.2010.521048

Witoon, T. (2011). Characterization of calcium oxide derived from waste eggshell and its application as CO2 sorbent. *Ceramic International, 37*(8), 3291–3298. https://doi.org/10.1016/j.ceramint.2011.05.125

Wu, J., Wei, H., & Peng, L. (2019). Research on the evolution of building technology based on regional revitalization. *Buildings, 9*(7), 165. https://doi.org/10.3390/buildings9070165

Yaacob, I. I., Ali, M. Y., Sopyan, L., & Hashmi, S. (2015). Effect of calcium carbonate replacement on workability and mechanical strength of Portland cement concrete. *Advanced Materials Research, 1115*(1), 131–137. https://doi.org/10.4028/www.scientific.net/AMR.1115.137

Zakaria, F. Z., Mihaly, J., Sajo, I., Katona, R., Hajba, L., Aziz, F. A., & Mink, J. (2008). FT-Raman and FTIR spectroscopic characterization of biogenic carbonates from Philippine venus seashell and Porites sp. Coral. *Journal of Raman Spectroscopy, 39*(9), 1204–1209. https://doi.org/10.1002/jrs.1964

Zhang, W., Zakaria, M., & Hama, Y. (2013). Influence of aggregate materials characteristics on the drying shrinkage properties of mortar and concrete. *Construction and Building Materials, 49*(1), 500–510. https://doi.org/10.1016/j.conbuildmat.2013.08.069

Fly Ash High Volume Concrete Cast with Plastic Waste Filler

Chin Mei Yun, Md Rezaur Rahman, Kuok King Kuok, Amelia Chai Pei Sze, Winston Wong Wen Ang, Muhammad Khusairy Bin Bakri, and Mohammed Mahbubul Matin

Abstract The workability and strength of green-mix concrete in an experiment using high fly ash concrete cast with various percentages of plastic trash are presented in this study. The specimens were cast into cubes and cylinders and cured for 7 days, 28 days, and 56 days in a water tank. Plastic waste fibres were added to the fly ash concrete with an aspect ratio of 15 at varied percentages of 0.5, 1.0, and 2.0%. The water-to-cement ratio was controlled at 0.5. The strength characteristics of standard concrete were compared to those of green-mix concrete. Control concrete using Ordinary Portland Cement (OPC) Concrete and High Volume Fly Ash (HVFA 40) Concrete cast at a set percentage of 40% in this experiment were used to compare the strength results of control concrete with control concrete HVFA 40 specimens containing varying percentages of plastic waste fibres. Cement was substituted with high volume fly ash at a constant proportion of 40% since, according to past research, that was the percentage that produced the best results in terms of strength. The strength of fly ash concrete without plastic waste fibre was much lower than that of regular concrete. The strength results improved when different percentages of plastic waste fibers were added to the green-mix concrete. However, after 56 days of curing, green-mix concrete with a constant of 40% fly ash replacement by additional 2% plastic fibres outperformed regular concrete in terms of strength.

Keywords Fly ash · Plastic waste fibres (PET) · Concrete · Workability · Strength

C. M. Yun · K. K. Kuok · A. C. P. Sze · W. W. W. Ang
Faculty of Engineering, Computing and Science, Swinburne University of Technology Sarawak Campus, Jalan Simpang Tiga, 93400 Kuching, Sarawak, Malaysia

M. R. Rahman (✉) · M. K. B. Bakri
Faculty of Engineering, Universiti Malaysia Sarawak, Jalan Datuk Mohammad Musa, 94300 Kota Samarahan, Sarawak, Malaysia
e-mail: rmrezaur@unimas.my

M. M. Matin
Department of Chemistry, Faculty of Science, University of Chittagong, Chittagong 4331, Bangladesh

M. R. Rahman et al. (eds.), *Waste Materials in Advanced Sustainable Concrete*, Engineering Materials, https://doi.org/10.1007/978-3-030-98812-8_5

1 Introduction

Concrete is one of the most widely utilized building materials on the planet (Gagg, 2014). Ordinary Portland Cement, water, coarse and fine particles are the essential components of concrete (Shah et al., 2020; Wang et al., 2020). Cement manufacturing necessitates a considerable quantity of raw materials, energy, and heat use, all of which emit gaseous emissions into the atmosphere (Ali et al., 2015; Bakhtyar et al., 2017; Worrell et al., 2001). Because the characteristics of fly ash and Portland cement powder are comparable, most cement producers have switched to using fly ash instead of Portland cement powder to reduce the environmental effect (Karasin & Dogruyol, 2014; Teara et al., 2019; Thomas, 2007).

Thermal power stations and coal-fired power plants both create fly ash. The ash is produced when coal is crushed in the boiler operation (Marinina et al., 2021). Bottom ash is the ash that forms below the boiler, whereas fly ash is the finer ash that forms above the boiler (Conklin, 1993; Marinina et al., 2021). Fly ash has been divided into two categories: Class C and Class F (Behera, 2010; Jais et al., 2019; Latifee, 2016; Saha, 2018; Thomas, 2007). The main distinction between these two types of fly ash is the chemical composition (Bhatt et al., 2019; Chelberg, 2019; Sola et al., 2011). Class F fly ash has pozzolanic qualities with less than 8% calcium oxide (CaO) inclusion, whereas Class C had pozzolanic properties with more than 8% calcium oxide (CaO) inclusion (Ramadevi and Manu, 2012). Due to the delayed pozzolanic reaction of fly ash, many researchers found that high volume fly ash (HVFA) concrete required additional setting time to acquire the strength characteristics (Bentz et al., 2013; Fan et al., 2015).

Diyora et al. (2014) studied how the partial substitution of fly ash influenced the strength characteristics of HVFA concrete. According to the studies, the optimal replacement percentage for achieving the best strength outcomes was 40% (Rehman, 2014). Yeh (1998) and Sravana et al. (2013) also looked at how the strength characteristics of HVFA concrete were impacted by different water to binder ratios. According to a few researchers, a low water-binder ratio increases the compressive strength of HVFA concrete because fly ash requires less water to promote cohesiveness while decreasing segregation (Asha & Reshmi, 2016; Herath et al., 2020; Khan et al. 2019; Moffat et al., 2017).

Concrete is widely used in construction because of its high compatibility and durability. It has low tensile and flexural bending strengths, which are needed to reinforce materials (Abbas et al., 2016; Naaman, 2001; Patrick et al., 2019). Prahallada and Prakash (2013) showed that polyethylene terephthalate (PET) fibre reinforced concrete might be used in construction due to its mechanical qualities and performance characteristics. The mechanical properties and performance characteristics of polyethylene terephthalate (PET) Fibers reinforced concrete have been demonstrated by Prahallada and Prakash (2013), allowing it to develop as a construction material due to its mechanical properties and performance characteristics when compared to ordinary concrete. According to Nibudey et al. (2013) and Prahallada and Prakash (2013), the use of PET fibres in concrete enhanced the strength characteristics compared to OPC concrete. Much research has been done on the effects of hardened elements on OPC and HVFA concrete (Huang et al., 2013; Yoo et al., 2017; @@Zhang et al., 2020). As a result, the percentage of fly ash replacement was also studied, and the most appropriate percentage of replacement fly ash was found to be 40% (Bahedh & Jaafar, 2018; Chen et al., 2019; Roshani et al., 2021; Zeggar et al., 2019). Apart from that, Ramadevi and Manju (2012) discovered that the PET fibres with an extra 2% had the maximum compressive and tensile strength compared to OPC concrete at the 28th day of curing age. But there is a gap of information that there is no research on the strength properties related to the addition of PET fibres in partial replacement of fly ash concrete. This study aimed to investigate the strength properties of the added different percentages of the PET fibres into 40% replacement of HVFA concrete while comparing the strength properties between OPC concrete and the HVFA concrete without adding PET fibres.

2 Methodology

2.1 *Materials*

Sejingkat Coal Power Plant in Kuching, Sarawak, provided low calcium fly ash (Class F) with a specific gravity of 2.4. Cahaya Mata Sarawak (CMS) Berhad's Ordinary Portland Cement (OPC) was utilized in this investigation. The OPC used to have a specific gravity of 3.15. Kasuma Engineering Sdn. Bhd. supplied the local river sand and gravels for the project. The sizes of the sand and gravels with a specific gravity of 2.6 range were 150–5 mm and 5–20 mm, respectively. The grading curve of the sand and particle size distribution (PSD) of the gravels used for concrete casting was designed according to the BS 12620:2002 (2002). The recycled PET plastic waste was obtained from local restaurants. The plastics were cut into 30 mm × 2 mm with an aspect ratio of 15. The various volume fractions of 0.5%, 1.0% and 2.0% plastic were used in the experiment.

2.2 *Mix Proportion and Sample Preparation*

The concrete samples were produced with a design mix ratio of 1:1.5:3 (cement, sand, and coarse aggregates) and fixed water to cement ratio of 0.5, corresponding to the M20 concrete grade. The concrete mixture method was produced was under the IS standards. The cement powder was replaced with a controlled amount of 40% fly ash, and different addition of percentages of plastic waste fibres, i.e., 0.5, 1.0, and 2.0%, were employed in the experiment. During the dry mix of concrete, plastic waste fibres cut to a consistent size of 30 mm × 2 mm and with an aspect ratio of 15 were added. The specimens were cast in 100 mm × 100 mm × 100 mm cube moulds for compressive strength and 100 mm diameter and 200 mm cylinder moulds for tensile strength. After even mixing, three layers of new concrete were poured into the cube and cylinder moulds, each receiving 25 blows. Before demolding for curing, the sample was left at room temperature for at least 24 h. Subsequently, the demoulded pieces were placed in water curing for either 7, 28, or 56 days before testing was carried out.

2.2.1 Workability Test

In the casting process, OPC, river sand as fine aggregates, and gravels with and without recycled PET plastic were evenly mixed using a concrete mixing drum. The water was added to the mixture, and the mixing continued for 15 min. The mixture was separated to run the slump cone test for the concrete workability according to ASTM C143/C143M-20 (2020).

2.3 *Compressive Strength Test*

The compressive strength test was carried out using a Unit Test Autocon 2000 (Promat (HK) Ltd., Hong Kong) in line with BS EN 12390-3 (2019). During the compressive strength test, the loads gradually increased to the samples at 2.4 ± 0.2 kN/s until the specimens failed. The results were averaged and recorded using concrete examples from the same batch.

2.4 *Tensile Strength Test*

The tensile strength test was carried out using a Unit Test Autocon 2000 (Promat (HK) Ltd., Hong Kong) under BS EN 12390-6 (2009). A cylinder sample was used for the determination of tensile strength. The results were recorded and averaged to represent the tensile strength of the batch.

2.5 *Scanning Electron Microscopy (SEM) and Energy Dispersive X-Ray/Spectroscopy (EDS/EDX)*

Scanning electron microscopy SEM and energy dispersive x-ray/spectroscopy (EDX/EDS) were performed in accordance with ASTM C1723-16 (2016) and ASTM

E1508-12 (2019) standards, respectively. The samples were examined at magnifications of 100x, 250x, 500x, and 10,000x. The elemental composition percentages of the samples are scanned and analyzed by automated software. The EDX/EDS was performed many times on each sample at various stages, with the most representative results chosen. For the materials' SEM and EDX/EDS examinations, Hitachi Ltd., Tokyo, Japan, utilized a Hitachi TM4000Plus Tabletop Microscope with a Quantax75TM Series Energy Dispersive X-Ray Spectrometer (Hitachi Ltd., Tokyo, Japan).

2.5.1 Fourier Transform Infrared Spectroscopy

Fourier-transform infrared spectroscopy (IRAffinity-1, Shimadzu Corporation, Kyoto, Japan) was used for the FTIR analysis of the samples. Fourier-transform infrared spectroscopy was conducted according to the ASTM E168-16 (2016) and ASTM E1252-98 (2013) standards for qualitative and quantitative analysis. The spectrum scanning was conducted in the wavenumber range of 4000–400 cm^{-1} for each sample. Fourier-transform infrared spectroscopy utilized the samples' infrared spectrum transmittance and absorption to develop a unique molecular fingerprint spectrum. The test was repeated numerous times for each sample, and the most representative results were selected.

3 Results and Discussions

3.1 Effect of HVFA and Recyeld Pet Plastics on the Workability of Mix Proportion

When Ordinary Portland cement was substituted with fly ash powder, the slump test results increased. The findings were backed up by Fan et al. (2015) and Singh et al. (2015). The delayed response of pozzolanic led the setting time of the fly ash to take considerably longer in the cement hydration process to establish the bonding, which enhanced the workability of HVFA concrete. When recycled PET plastic was added, the slump of the fly ash concrete sample was reduced dramatically. In other words, as the amount of recycled PET plastic in the concrete mixture rose, the workability of the mix declined. The other recycled PET plastic increases the total surface area, but water content remains the same, thus reducing the workability (Kassa et al., 2019). Besides, the interlocking of long strip PET fibres increases the resistance in the concrete mix hence affecting the workability (Shamsudin et al., 2021). The form and size of the recycled PET plastics, which impede movement within the mixture, was the lower workability of concrete mixtures (Al-Hadithi & Abbas, 2018).

3.2 Compressive Strength of Concrete Samples

Figure 1 shows the compressive strength tests performed on OPC and HVFA40 concrete with various percentages of recycled PET plastics for the curing sample of 7, 28, and 56 days.

Compared to the HVFA 40 concrete added with 0.5, 1, and 2 recycled PET plastic, the HVFA 40 concrete without any recycled PET plastic (15.79 MPa) had the lowest compressive strength values at 7 days of curing age (16.52, 16.26, 19.7 MPa). The inclusion of recycled PET plastic enhanced the compressive strength of the HVFA 40 concrete, but it was still unable to surpass the compressive strength of OPC concrete (27.18 MPa) at 7 days. When recycled PET plastic was added into HVFA 40 concrete, the early age compressive strength did not improve much for 0.5 and 1% addition. The reason was that the incomplete hydration of binder materials, primarily fly ash, has a retarding effect on strength development, resulting in poor bonding of concrete matrix.

The 28-day compressive strength of HVFA 40 concrete increases with additional recycled PET plastic. According to the findings, the compressive strength of HVFA 40 concrete with 0.5, 1.0, and 2.0% recycled PET plastic was unable to exceed that of OPC concrete (29.53 MPa), but the HVFA 40 with 2% recycled PET plastic (27.72 MPa) showed a very close match to OPC concrete at 28 days. From 7 to 28 days, they added recycled PET plastic to concrete, substantially enhancing the compressive strength. It has been demonstrated that recycled PET plastic may be used as strengthening materials with interlocking properties to increase the strength of concrete between aggregates (Prahallada & Prakash, 2013). The addition of recycled

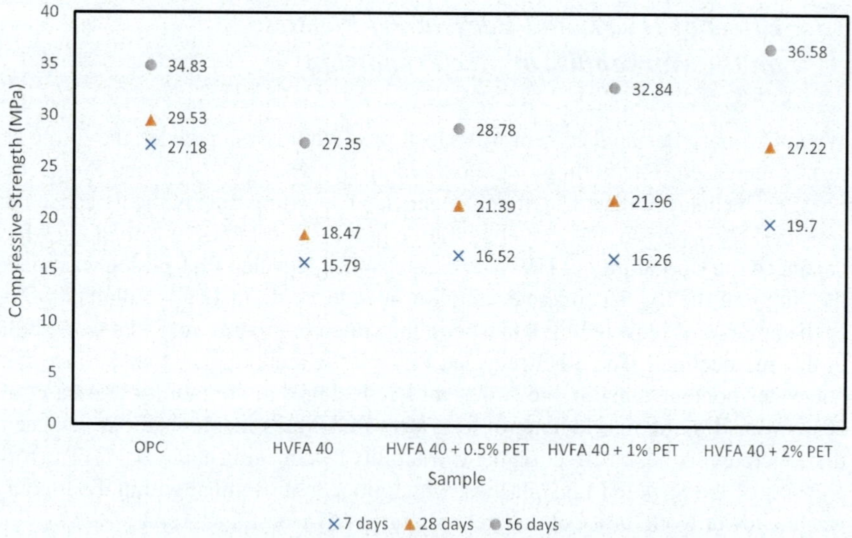

Fig. 1 Compressive strength with curing age of 7, 28 and 56 days

PET plastic, on the other hand, may have enhanced the absorbed energy before failure, increasing compressive strength (Al-Hadithi & Abbas, 2018). However, the results showed that HVFA 40 concrete specimens containing 2% recycled PET plastic could outperform OPC concrete in compressive strength after 56 days of curing age. It was not due to recycled PET plastic but rather to the fly ash itself.

At 56 days, the HVFA 40 concrete specimens with 2% recycled PET plastic has average compressive strength (36.58 MPa) that surpass OPC concrete (34.83 MPa). As 2% recycled PET plastic was added to the specimens, the compressive strength increased significantly compared to HVAF 40 with 0.5% and 1% recycled PET plastic. Adedapo (2007), Patil and Patil (2017), and Ramadevi and Manju (2012) conducted another experiment using HVFA 40 concrete specimens, including 2% recycled PET plastic, which demonstrated to be the optimal proportion for obtaining the maximum compressive strength values. On the other hand, it was discovered that the compressive strength of HVFA 40 concrete continues to increase after the 28th day of hardening. The findings of Diyora et al. (2015) show that due to the delayed pozzolanic reaction, fly ash took a considerably longer time to generate compressive strength.

Aside from the percentage of recycled PET plastic added into the concrete mixture, other factors that can affect the compressive strength of concrete was the shape and aspect ratio of recycled PET plastic. Asha and Reshmi (2016) found that the optimum aspect ratio was 15 in OPC concrete. On the other hand, Borg et al. (2016) found that recycling PET plastic with an aspect ratio of 15 produces OPC concrete with higher compressive than recycling plastic with an aspect ratio of 25. Besides, straight recycled PET plastic produces lower compressive strength than crinkle recycled PET plastic.

3.3 *Tensile Strength of Concrete Samples*

The tensile strength tests were conducted between OPC concrete and HVFA concrete by adding varying percentages of recycled PET plastic samples and tested at varying curing ages ranging from 7 to 56 days. Figure 2 shows the results. Compared to HVFA 40 concrete with 0.5%, 1%, and 2% recycled PET plastic (8.19, 7.17, 8.21 MPa), the tensile strength of HVFA 40 concrete without any recycled PET plastic (7 MPa) at 7 days showed the lowest compressive strength. Although the addition of recycled PET plastic to HVFA 40 concrete enhanced tensile strength, it can still not match the strength of OPC concrete after 7 days of curing (11.38 MPa).

Compared to HVFA 40 concrete with 0.5 and 1.0% recycled PET plastic, the HVFA 40 concrete with 2% recycled PET plastic had the most excellent tensile strength result (12.62 MPa) at 28 days. The addition of recycled PET plastic to concrete enhanced the tensile strength dramatically from 7 to 28 days. However, at 28 days, the HVFA 40 concrete containing 2% recycled PET plastic (12.62 MPa) demonstrated tensile strength findings higher than the OPC concrete (11.52 MPa).

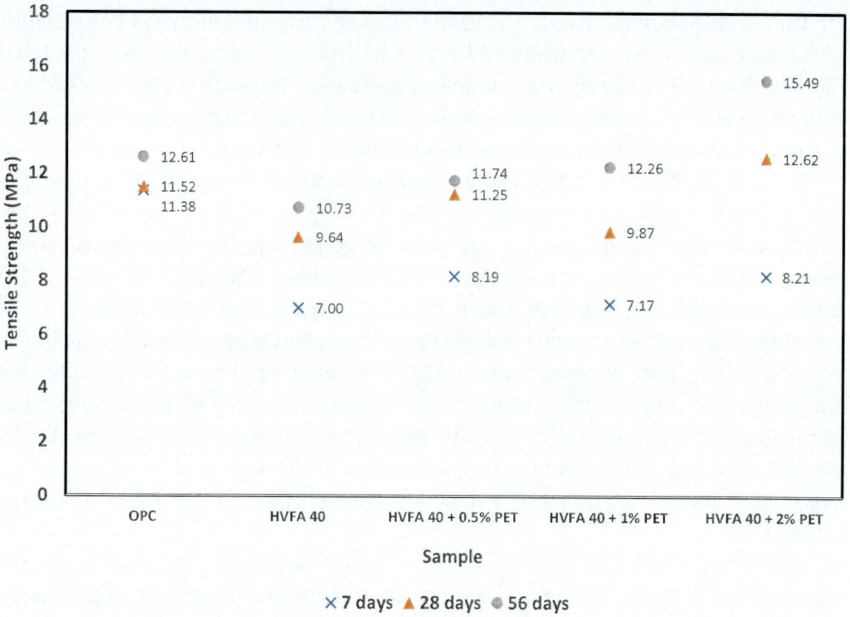

Fig. 2 Tensile strength with curing age of 7, 28 and 56 days

Singh and Khandelwal's (2020) study showed that the optimum amount of recycled PET plastic fibres with an aspect ratio of 45 in fly ash and ground granulated blast furnace slag geopolymer concrete was 3%; a decrease in tensile strength was observed when recycling PET plastics fibres was raised to 4%. While Asha and Reshmi (2016) found that adding 1% recycle PET plastic into the concrete mixture resulted in the highest tensile strength. On the other hand, aspect ratio 15 gives the best result in tensile strength compared to aspect ratios 8 and 23. Also, the crinkled shaped recycled PET plastic fibres have better tensile strength than straight shaped recycled PET plastic fibres.

Due to the delayed functionalization of the bonding effect via the hydration process, the recycled PET plastic functioned as strengthening materials in HVFA 40 concrete throughout the curing age, resulting in enhanced tensile strength (Al-Hadithi & Abbas, 2018). According to Patil and Patil (2017), specimens with additional recycled PET plastic fibers of 1–2% showed a rise in tensile strength. Furthermore, because of the delayed pozzolanic reaction, the rise in tensile strength was driven by fly ash, which takes a considerably longer time to generate compressive strength (Diyora et al. 2015). Another experiment conducted by Diyora et al. (2015) revealed that the HVFA 40 concrete continued to increase in strength between the 28th and 91st days of curing due to the bonding effect's contribution. Due of a time constraint, the experiment was not carried out after 91 days of curing. However, at 56 days, tensile strength tests revealed that HVFA 40 concrete with 2% recycled PET plastic outperformed OPC concrete.

3.4 Morphological Properties

The SEM images for HVFA 40 + 2% PET (7 days) are shown in Fig. 3a–d. Figure 3c, d show that the fly ash structure is spherical due to the ball milling and sieving method used in fine fly ash manufacturing. In Fig. 3c, d, there were several cenospheres structures (expended mineral materials, including alumina and silica), pores/void on the sphere, fine and irregular shaped of plastic and fly ash. It might be the result of an unrequited chemical interaction between various components. However, as seen in Fig. 3d, part of the fly ash and plastic barely maintain their typical spherical and original shapes, respectively (Wang et al., 2008). On the other hand, it should be noted that not all spherical fly ash particles and plastic have a rough pore surface, and a minority of them have a smooth surface, which prevents appropriate mixture bonding.

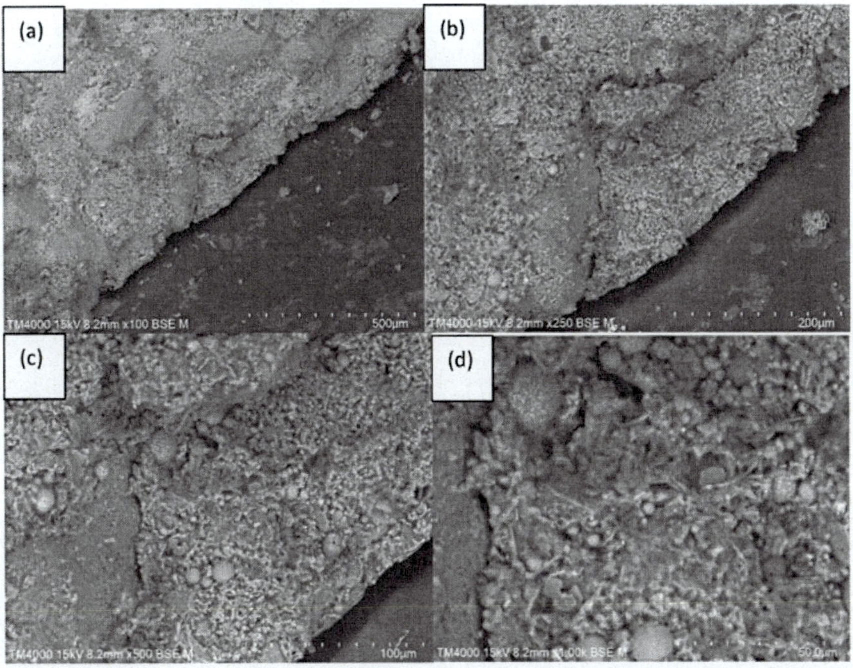

Fig. 3 SEM images for HVFA 40 + 2%PET (7 days) at magnification of **a** 100x, **b** 250x, **c** 500x, and **d** 1000x

3.5 Elemental Analysis

Figure 4 illustrates the spectrum of the association with the data in Table 1, while Table 1 gives the EDS/EDX element contents HVFA 40 + 2%PET (7 days). Table 1 shows that gold (Au), carbon (C), oxygen (O), calcium (Ca), silicon (Si), and aluminium (Al) are all present (Al). Carbon had the most significant atom percentage, oxygen and gold, while the other elements had low mass percentages, as seen in Table 1. It shows that there is a possibility of forming SiO_2, $CaCO_3$, Al_20_3, CaO, and MgO. The presence of gold might be due to fly ash content defects, especially traces of e-waste. Fly ash (Class F) has a greater Si and Al concentration and a more alumino-silicate amorphous structure. This unstructured $Ca(OH)_2$ interacts with cement hydration processes to form calcium-aluminate-hydrates (Ca-Al-H) and calcium-silicate-hydrates (Ca–Si-H) (Fauzi et al., 2016).

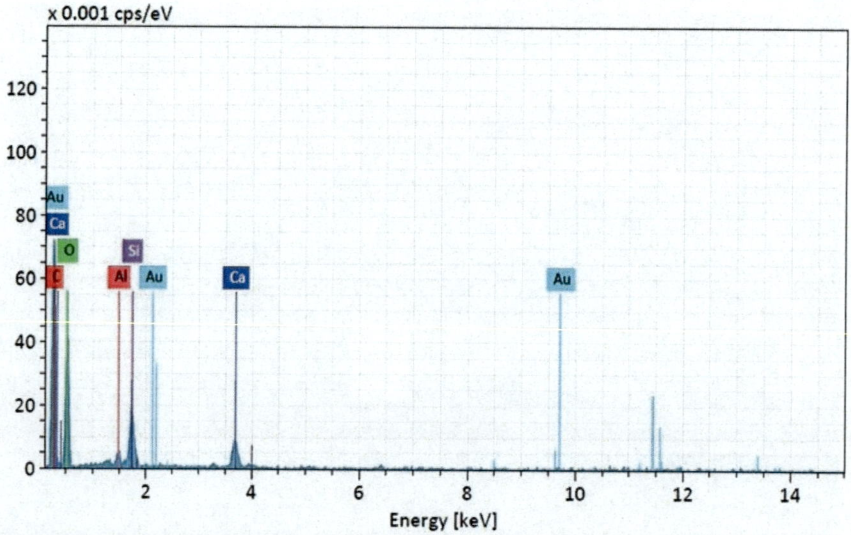

Fig. 4 EDS/EDX graph for HVFA 40 + 2%PET (7 days)

Table 1 Element composition of HVFA 40 + 2%PET (7 days)

Element	Mass normal (%)	Atom (%)
Au	84.39	27.13
C	10.15	53.52
O	4.41	17.44
Ca	0.70	1.11
Si	0.30	0.67
Al	0.06	0.13
Sum	100.00	100.00

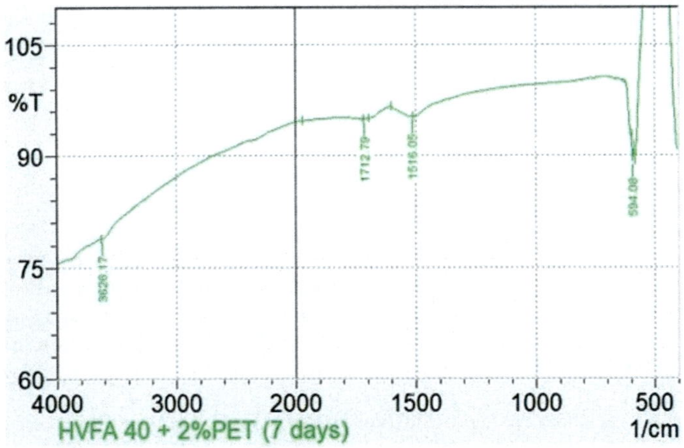

Fig. 5 FTIR spectra of HVFA 40 + 2%PET (7 days)

3.6 Infrared Spectral Properties

The FTIR pictures for HVFA 40 + 2%PET (7 days) are shown in Fig. 5. At a band intensity of 3500–3000 cm^{-1}, the –OH stretching and H–O–H vibrations bending is revealed. In contrast, it is indicated for water molecules absorbed and entrapped in porous materials. Because most of the water has been hydrolyzed, the peak is narrower. 1712.79 and 1516.05 cm^{-1} are the broad bands recommended for Si–O–Si bond bending and stretching, respectively (Fauzi et al., 2016; Garcia et al. 2009; Puligilla & Mondal, 2015). Finally, the bend O–Si–O and Si–O–Si bonds have less than 594.08 cm^{-1} bands (Fauzi et al., 2016; Garcia et al., 2009; Puligilla & Mondal, 2015).

4 Conclusion

Finally, the HVFA concrete was used in this study, which was mixed with various percentages of recycled PET plastic. As the cement powder was substituted with high-volume fly ash powder, the workability of the mix concrete improved. As the amount of recycled PET plastic added rose, the workability of the green-mix concrete decreased. Due to the contribution of the bonding effect, the increase in strength characteristics of green-mix concrete is more substantial after 28 days of curing. As the amount of recycled PET plastic in green-mix concrete was raised, the strength characteristics of the concrete improved considerably. After 56 days of curing, mixed concrete had the highest compressive and tensile strength values compared to OPC concrete. The optimal proportion of recycled PET plastic to add to HVFA 40 concrete to get the highest compressive and tensile strength values is 2%.

Acknowledgements The authors would like to thank Swinburne University of Technology Sarawak Campus and Universiti Malaysia Sarawak (UNIMAS) for the collaboration.

References

Abbas, S., Mehdi, M. L., & Saleem, M. A. (2016). Ultra-high performance concrete: mechanical performance, durability, sustainability and implementation challenges. *International Journal of Concrete Structures and Materials, 10*(1), 271–295. https://doi.org/10.1007/s40069-016-0157-4

Adedapo, A. A. (2007). Pavement deterioration and PE pipe behaviour resulting from open-cut and HDD pipeline installation techniques. *Ph.D. thesis*, University of Waterloo (Canada), Ann Arbor, p. 412.

Al-Hadithi, A. I., & Abbas, M. A. (2018). The effects of adding waste plastic fibers on the mechanical properties and shear strength of reinforced concrete beams. *Iraqi Journal of Civil Engineering, 12*(1), 110–124.

Ali, N., Abbas, J., Answer, M., Alwi, S. K. K., & Anjum, M. N. (2015). The greenhouse gas emissions produced by cement production and its impact on environment: a review of global cement processing. *International Journal of Research, 2*(2), 488–500.

Asha, S., & Reshmi, P. R. (2016). Use of post consumed straight and crimped waste plastic fibres in concrete. *IOSR Journal of Mechanical and Civil Engineering, 20*(201), 32–41. https://www.iosrjournals.org/iosr-jmce/papers/ICETEM/Vol.%201%20Issue%201/06-32-41.pdf

ASTM E1508–12. (2019). *Standard guide for quantitative analysis by energy-dispersive spectroscopy.* ASTM International. https://doi.org/10.1520/E1508-12AR19

ASTM C1723-16. (2016). *Standard guide for examination of hardened concrete using scanning electron microscopy.* West Conshohocken: ASTM International. https://doi.org/10.1520/C1723-16

ASTM E168-16. (2016). *Standard practices for general techniques of infrared quantitative analysis.* ASTM International. https://doi.org/10.1520/E0168-16

ASTM C143/C143M-20. (2020). *Standard test method for slump of hydraulic-cement concrete.* West Conshohocken: ASTM International. https://doi.org/10.1520/C0143_C0143M-20

Bahedh, M. A., & Jaafar, M. S. (2018). Ultra high-performance concrete utilizing fly ash as cement replacement under autoclaving technique. *Case Studies in Construction Materials, 9*(1), e00202. https://doi.org/10.1016/j.cscm.2018.e00202

Bakhtyar, B., Kacemi, T., & Nawaz, M. A. (2017). A review on carbon emissions in Malaysia cement industry. *International Journal of Energy Economics and Policy, 7*(3), 282–286.

Behera, R. K. (2010). Characterization of fly ash for their effective management and utilization. *Degree Thesis*, National Institute of Technology Rourkela (pp.1–61)

Bentz, D. P., Ferraris, C. F., & Snyder, K. A. (2013). Best practices guide for high-volume fly ash concretes: assuring properties and performance. *NIST Technical Note, 1812*, 1–56. https://doi.org/10.6028/NIST.TN.1812

Bhatt, A., Priyadarshini, S., Mohanakrishnan, A. A., Abri, A., Sattler, M., & Techapaphawit, S. (2019). Physical, chemical, and geotechnical properties of coal fly ash: a global review. *Case Studies in Construction Materials, 11*(1), e00263. https://doi.org/10.1016/j.cscm.2019.e00263

Borg, R. P., Baldacchino, O., & Ferrara, L. (2016). Early age performance and mechanical characteristics of recycled PET fibre reinforced concrete. *Construction and Building Materials, 108*, 29–47. https://doi.org/10.1016/j.conbuildmat.2016.01.029

BS 12620:2002. (2002). *Aggregate for concrete.* British Standard

BS EN 12390-3. (2019). *Testing hardened concrete. Compressive strength of test specimens.* British Standard Institution.

BS EN 12390-6. (2009) *Testing hardened concrete. Tensile splitting strength of test specimens.* British Standard Institution.

Chelberg, M. (2019). The effect of fly ash chemical composition on compressive strength of fly ash portland cement concrete. Master Thesis. The Ohio State University (pp. 1–64).

Chen, H.-J., Shih, N.-H., Wu, C.-H., & Lin, S.-K. (2019). Effects of the loss on ignition of fly ash on the properties of high-volume fly ash concrete. *Sustainability, 11*(9), 2704. https://doi.org/10.3390/su11092704

Conklin, T.L. (1993). Utilization of fly ash in highway construction, *Master thesis*. University of Wyoming (p. 91).

Diyora, S., Solanki, G., & Chandiwala, A. K. (2014). A Review on compressive strength of plain OPC cement concrete with fly ash using different water-cement ratio. *International Journal of Advance Engineering and Research Development, 1*(12), 129–134.

Fan, W.-J., Wang, X.-Y., & Park, K.-B. (2015). Evaluation of the chemical and mechanical properties of hardening high-calcium fly ash blended concrete. *Materials, 8*(9), 5933–5952. https://doi.org/10.3390/ma8095282

Fauzi, A., Nuruddin, M. F., Malkawi, A. B., & Abdullah, M. M. A. B. (2016). Study of fly ash characterization as a cementitious material. *Procedia Engineering, 148*(1), 487–493. https://doi.org/10.1016/j.proeng.2016.06.535

Gagg, C. R. (2014). Cement and concrete as an engineering material: an historic appraisal and case study analysis. *Engineering Failure Analysis, 40*(1), 114–140. https://doi.org/10.1016/j.engfailanal.2014.02.004

García Lodeiro, I., Macphee, D. E., Palomo, A., & Fermandez-Jimenez, A. (2009). Effect of alkalis on fresh C–S–H gels. FTIR analysis. *Cement and Concrete Research, 39*(3), 147–153. https://doi.org/10.1016/j.cemconres.2009.01.003

Herath, C., Gunasekara, C., Law, D. W., & Setunge, S. (2020). Performance of high volume fly ash concrete incorporating additives: a systematic literature review. *Construction and Building Materials, 258*(1), 120606. https://doi.org/10.1016/j.conbuildmat.2020.120606

Huang, C.-H., Lin, S.-K., Chng, C.-S., & Chen, H.-J. (2013). Mix proportions and mechanical properties of concrete containing very high-volume of Class F fly ash. *Construction and Building, 46*(1), 71–78. https://doi.org/10.1016/j.conbuildmat.2013.04.016

IS 10262. (2019). Concrete mix proportioning—Guidelines. Indian Standard.

IS 456. (2000). *Plain and reinforced concrete—Code practice.* Indian Standard.

Jais, I. B. M., Abdullah, N., Ali, M. A. M., & Johar, M. A. (2019). Peat modification integrating geopolymer and fly ash. *IOP Conference Series: Materials Science and Engineering, 527*(1), 012021. https://doi.org/10.1088/1757-899X/527/1/012021

Karasin, A., & Dogruyol, M. (2014). (2014) An experimental study on strength and durability for utilization of fly ash in concrete mix. *Advances in Materials Science and Engineering, 1*, 1–6. https://doi.org/10.1155/2014/417514

Kassa, R. B., Kanali, C., & Ambassah, N. (2019). Engineering properties of polyethylene terephthalate fibre reinforced concrete with fly ash as a partial cement replacement. *Civil and Environmental Research, 11*(6). https://doi.org/10.7176/CER

Khan, M. S., Shariq, M., Akhtar, S., & Masood, A. (2019). Performance of high-volume fly ash concrete after exposure to elevated temperature. *Journal of the Australian Ceramic Society, 56*(1), 781–794. https://doi.org/10.1007/s41779-019-00396-6

Latifee, E. R. (2016). An experimental factorial design for ASR mitigation with fly ash. *Malaysian Journal of Civil Engineering, 28*(1), 124–138. https://engineering.utm.my/civil/mjce/wp-content/uploads/sites/40/2016/04/Vol-281-Paper-9.pdf

Marinina, O., Nevskaya, M., Jonek-Kowalska, I., Wolniak, R., & Marinin, M. (2021). Recycling of coal fly ash as an example of an efficient circular economy: a stakeholder approach. *Energies, 14*(12), 3597. https://doi.org/10.3390/en14123597

Moffat, E. G., Thomas, M. D. A., & Fahim, A. (2017). Performance of high-volume fly ash concrete in marine environment. *Cement and Concrete Research, 102*(1), 127–135. https://doi.org/10.1016/j.cemconres.2017.09.008

Naaman, A. E. (2001). Reinforced concrete. In: Buschow, K. H. J., Cahn, R. W., Flemings, M. C., Ilschner, B., Kramer, E. J., Mahajan, S., Veyssiere, P. (Eds.) *Encyclopedia of materials: Science and technology*. Pergamon (pp. 8095–8109). https://doi.org/10.1016/B0-08-043152-6/01454-6

Nibudey, R. N., Nagarnaik, P. B., Parbat, D. K., & Pande, A. M. (2013). A model for compressive strength of PET fiber reinforced concrete. *American Journal of Engineering Research, 2*(12), 367–372.

Patil, V., & Patil, R. (2017). Compressive and tensile strength analysis of polyethylene terephthalate (PET) for 7, 28 and 56 days. *International Research Journal of Engineering and Technology, 4*(12), 351–353. https://www.irjet.net/archives/V4/i12/IRJET-V4I1265.pdf

Patrick, K., Onchiri, R. O., & Mang'uriu, G. N. (2019). Developing suitable proportions for the production of pineapple leaf fibers reinforced normal strength concrete. *Open Journal of Civil Engineering, 9*(3), 185–194. https://doi.org/10.4236/ojce.2019.93013

Prahallada, M. C., & Prakash, K. B. (2013). Bond characteristics of waste plastic fibre reinforced concrete - an experimental investigation. *I-Manager's Journal on Structural Engineering, 2*(2), 30–33.

Puligilla, S., & Mondal, P. (2015). Co-existence of aluminosilicate and calcium silicate gel characterized through selective dissolution and FTIR spectral subtraction. *Cement and Concrete Research, 70*(1), 39–49. https://doi.org/10.1016/j.cemconres.2015.01.006

Ramadevi, K., & Manju, R. (2012). Experimental Investigation on the properties of concrete with plastic pet(bottle) fibres as fine aggregates. *International Journal of Emerging Technology and Advanced Engineering, 2*(6), 42–46.

Rehman, S. (2014). The effect of the heavy metal immobilization agents on the strength of fly ash treated soils and the long-term stability of the agents. *Master Thesis*, The University of Texas at San Antonio, Ann Arbor, p. 87.

Roshani, M. M., Kargar, S. H., Farhangi, V., & Karakouzian, M. (2021). Predicting the effect of fly ash on concrete's mechanical properties by ANN. *Sustainability, 13*(3), 1469. https://doi.org/10.3390/su13031469

Saha, A. K. (2018). Effect of class F fly ash on the durability properties of concrete. *Sustainable Environment Research, 28*(1), 25–31. https://doi.org/10.1016/j.serj.2017.09.001

Shah, K. W., Huseien, G. F., & Xiong, T. (2020) Functional nanomaterials and their applications toward smart and green buildings. In: Samiu, P., Kim, D., Iyer, N.R., Chaudhary, S. (eds) *New materials in civil engineering*. Butterworth-Heinemann. https://doi.org/10.1016/B978-0-12-818961-0.00011-9

Shamsudin, M. M. H., Hamid, N. H., & Fauzi, M. A. M. (2021). Compressive and flexural strength of concrete containing recycled polyethylene terephthalate (PET). *Key Engineering Materials, 879 KEM*, 13–21. https://doi.org/10.4028/www.scientific.net/KEM.879.13

Singh, A., & Vikas Khandelwal, E. (2020). To study the effect of pet fiber (polyethylene terephthalate fiber) on geopolymer concrete by using GBBS and Fly Ash. *International Journal of Engineering Research & Technology, 9*(10), 170–177.

Singh, L. P., Goel, A., Bhattacharyya, S. K., Sharma, U., & Mishra, G. (2015). Hydration studies of cementitious material using silica nanoparticles. *Journal of Advanced Concrete Technology, 13*(7), 345–354. https://doi.org/10.3151/jact.13.345

Sola, O. C., Yayla, M., Sayin, B., & Atis, C. D. (2011). (2011) The effects of different types of fly ash on the compressive strength properties of briquettes. *Advances in Materials Science and Engineering, 1*, 1–6. https://doi.org/10.1155/2011/430604

Sravana, D., Sarika, P., Rao, S., Sekhtar, T. S., & Apparao, G. (2013). Studies on relationship between water/binder ratio and compressive strength of high volume fly ash concrete. *American Journal of Engineering Research, 2*(8), 115–122. http://www.ajer.org/papers/v2(8)/M028115122.pdf

Teara, A., Doh, S. I., Chin, S. C., Ding, Y. J., Wong, J., & Jiang, X. X. (2019). Investigation on the durability of use fly ash and eggshells powder to replace the cement in concrete productions. *National Colloquium on Wind & Earthquake Engineering, 244*(1), 012025. https://doi.org/10.1088/1755-1315/244/1/012025

Thomas, M. (2007). Optimizing the use of fly ash in concrete. *Concrete, 1*(1), 1–24. https://www.cement.org/docs/default-source/fc_concrete_technology/is548-optimizing-the-use-of-fly-ash-concrete.pdf

Wang, L., Chen, M. L., & Tsang, D. C. W. (2020). Green remediation by using low-carbon cement-based stabilization/solidification approaches. In: D. Hou (Ed.), *Sustainable remediation of contaminated soil and groundwater—materials, processes, and assessment* (pp. 93–118). Butterworth-Heinemann. https://doi.org/10.1016/B978-0-12-817982-6.00005-7

Wang, S., Baxter, L., & Fonseca, F. (2008). Biomass fly ash in concrete: SEM, EDX and ESEM Analysis. *Fuel, 87*(3), 372–379. https://doi.org/10.1016/j.fuel.2007.05.024

Worrell, E., Price, L., Maritin, N., Hendriks, C., & Media, L. O. (2001). Carbon dioxide emissions from the global cement industry. *Annual Review of Energy and the Environment, 26*(1), 303–329. https://doi.org/10.1146/annurev.energy.26.1.303

Yeh, I.-C. (1998). Modeling of strength of high-performance concrete using artificial neural networks. *Cement and Concrete Research, 12*(1), 1797–1808. https://doi.org/10.1016/S0008-8846(98)00165-3

Yoo, S.-W., Choi, Y. C., & Choi, W. (2017). (2017) Compression behavior of confined columns with high-volume fly ash concrete. *Advances in Materials Science and Engineering, 1*, 1–11. https://doi.org/10.1155/2017/8208079

Zeggar, M. L., Azline, N., & Safiee, N. A. (2019). Fly ash as supplementry material in concrete: a review. *IOP Conference Series: Earth and Environmental Science, 357*(1), 012025. https://doi.org/10.1088/1755-1315/357/1/012025

Zhang, P., Wang, L., Wei, H., & Wang, J. (2020) A critical review on effect of nanomaterials on workability and mechanical properties of high-performance concrete. *Advances in Civil Engineering, 2021*(1), 1–24. https://doi.org/10.1155/2021/8827124

Bottom Ash as Sand Filler Replacement in Concrete

Chin Mei Yun, Md Rezaur Rahman, Kuok King Kuok, Amelia Chai Pei Sze, Javan Liew San Jer, Muhammad Khusairy Bin Bakri, and Mohammed Mahbubul Matin

Abstract As the world's infrastructure develops, so does the need for concrete building materials. Malaysia is one of the nations that use pulverized coal to create power, and the burning of coal produces tons of bottom ash, which is considered industrial waste. Industrial waste disposal must be appropriately handled since it may influence the environment. Researchers have looked at the possibility of using this industrial waste as a component in concrete manufacturing to save natural resources. This work aims to investigate the characteristics of normal weight concrete using coal bottom ash (CBA) as a partial replacement for sand as fine aggregate. To make M30 grade concrete with a constant water-cement ratio, the replacement percentage of sand as fine aggregate with CBA utilized in this research project is 10, 20, 30, 40, and 50% by volume. The control concrete specimen was cast and tested with 100% sand as fine aggregate. There were 36 specimens cast and tested in total. The specimens were water cured for 7 and 28 days, following which their compressive strength was measured. Its workability and compressive strength test results were compared to standard concrete. CBA has shown encouraging results in compressive strength, solving the problem of industrial waste disposal, and preserving natural resources.

Keywords Normal concrete · Sand aggregate · Bottom ash · Compressive strength

C. M. Yun · K. K. Kuok · A. C. P. Sze · J. L. S. Jer
Faculty of Engineering, Computing and Science, Swinburne University of Technology, Sarawak Campus, Jalan Simpang Tiga, 93400 Kuching, Sarawak, Malaysia

M. R. Rahman (✉) · M. K. B. Bakri
Faculty of Engineering, Universiti Malaysia Sarawak, Jalan Datuk Mohammad Musa, 94300 Kota Samarahan, Sarawak, Malaysia
e-mail: rmrezaur@unimas.my

M. M. Matin
Department of Chemistry, Faculty of Science, University of Chittagong, Chittagong 4331, Bangladesh

M. R. Rahman et al. (eds.), *Waste Materials in Advanced Sustainable Concrete*, Engineering Materials, https://doi.org/10.1007/978-3-030-98812-8_6

1 Introduction

Due to its durability, concrete is the most widely utilized material in large building projects. The demand for building materials is rising in tandem with infrastructure growth worldwide. Cement and natural resources such as coarse aggregate and fine aggregates are the primary component elements of concrete (Singh & Siddique, 2015). Globally, around 2.7 billion m^3 of concrete was produced in 2002, with estimates that this figure will rise to over 7.5 billion m^3 yearly by 2050 (Rafieizonooz et al., 2016). Due to the massive volume of concrete used, it negatively impacts the environment, necessitating more sand aggregate usage. As a result, new alternative materials to replace sand aggregate in the manufacture of concrete are required. It also assists in reducing sand aggregate depletion, but it also provides a long-term future for the use of recycled products.

Coal bottom ash (CBA) is one by-product of pulverized coal-fired power production. Tanjung Bin power plant in Malaysia produces 180 tons of bottom ash and 1,620 tons of fly ash per day from 18,000 tons of coal burned daily (Abubakar & Baharudin, 2012). These by-products are regarded as waste and are disposed of in landfills. However, the disposal of CBA has reached a worrisome level, necessitating its re-use as a building material. Compared to river sand, CBA has a comparable look and particle size distribution (Singh and Siddique 2014a).

In comparison to natural sand, CBA is lighter and more brittle. Furthermore, the particles in CBA can interact with one another. Low specific gravity CBA has a porous texture that degrades faster when loaded or compacted. Because of its workability and strength, CBA is an excellent material to utilize as a partial replacement for sand aggregates (Singh & Siddique, 2016).

Over the years, there has been researched into using agricultural wastes as building materials in construction. The characteristics of concrete with varying percentages of sand aggregate replaced with CBA were examined and analyzed. Bottom ash has been used in the building sector for several purposes. Road foundation and sub-base, structural fill, backfill, drainage media, concrete aggregate, and manufactured soil products are just a few examples (Singh & Siddique, 2013). Replacing sand aggregate with bottom ash resulted in identical strength development as concrete without bottom ash. The compressive strength of concrete produced using bottom ash as a sand aggregate substitute was not significantly influenced (Kim & Lee, 2011).

According to Singh and Siddique (2014a), the workability of concrete containing bottom ash reduced when sand aggregate was partially or entirely replaced. The slump and compaction value of concrete fall as the percentage of bottom ash replaced increases. It is due to bottom ash porous particles, which have a greater water absorption rate than the sand aggregate. According to the study, the compressive strength of concrete with bottom ash diminishes after seven days. After 28 days of curing, the compressive strength of the concrete mixture rose and exceeded that of the control concrete mixture. It is due to bottom ash's delayed pozzolanic action.

According to the study, due to the high-water absorption of bottom ash, the workability of concrete decreases as the replacement percentage of bottom ash increases. Because the specific gravity of bottom ash is significantly lower than sand aggregate, the density of concrete containing bottom ash decreases as the amount of bottom ash increases. The researchers also tested the effect of different percentages of bottom ash substitution on concrete strength. In this experiment, the sand aggregate was substituted at 0, 10, 15, 20, and 25% for grade M30 concrete, with the findings indicating that the best replacement percentage of bottom ash was determined to be 15%. The findings of the compressive strength test after seven days show a decrease in strength. However, after 28 days of curing, concrete with a 15% replacement percentage had a strength value of 34.58 MPa, whereas regular concrete had a strength value of 31.70 MPa (Kajal et al., 2017).

According to a study, replacing 30% is optimum without adding superplasticizers when considering workability and strength characteristics. In contrast, a replacement percentage of up to 50% is advised when superplasticizers are added to the concrete (Singh & Siddique, 2016). The compressive strength of concrete improves significantly as the curing time increases (Singh & Siddique, 2014b). The compressive strength of concrete was evaluated at 0, 20, 30, 40, 50, 75, and 100% substitution of bottom ash (concrete mixture M0, M1, M2, M3, M4, M5 and M6). After seven days of curing, M2, M3, M4, M5, and M6 concrete mixtures were lower than M0 and M1. The concrete mixture has acquired compressive strength equivalent to the control concrete mixture as it has aged. The compressive strength of concrete mixture M3 (40% replacement) is 37.8 N/mm^2, close to the control mixture's 38.2 $^N/mm^2$.

This study's research gap has been discovered. The ideal proportion of bottom ash replacement in concrete differs from case study to case study, which is the research project's knowledge gap. As a result, this study aims to establish the optimal amount of bottom ash replacement in concrete by comparing the strength development to regular concrete.

1.1 Advantages and Disadvantages of Coal Bottom Ash

1.1.1 Coal Bottom Ash Properties

The quality of coal bottom ash was influenced by the type and quality of the coal used for combustion. Coal bottom ash from anthracite or bituminous coals has pozzolanic properties, while coal bottom ash from lignite or sub-bituminous coals has pozzolanic and cementitious properties. The pozzolanic properties was contribute from the high content of SiO_2, Fe_2O_3 and Al_2O_3 (Singh, 2018).

Coal bottom ash used to replace fine aggregate has its advantages and disadvantages. Compared to river sand or quarry sand, coal bottom ash has different chemical and physical properties. Generally, coal bottom ash has lower specific gravity and fineness modulus than the river sand but is higher in water absorption, as shown in Table 1.

Table 1 Properties of materials

Reference	Materials	Specific gravity	Water absorption (%)	Fineness modulus	SiO_2 (%)	Al_2O_3 (%)	Fe_2O_3 (%)	CaO (%)	MgO (%)	SO_3 (%)	Na_2O (%)	K_2O (%)	TiO_2 (%)
Singh and Siddique (2014a; b)	River sand	2.6	2.46	1.97									
	Bottom ash	1.39	31.58	1.37	47.53	20.69	5.99	4.17	0.82	1	0.33	0.76	-
Singh and Siddique (2016)	River sand			1.97									
	Quarry sand			2.58									
	Bottom ash	1.39	31.58	1.37	56.44	29.24	8.44	0.75	0.4	0.24	0.09	1.24	–
Andrade et al., (2009)	Cement	3.12			18.13	4.28	2.54	59.8	5.25	3.14	–	–	–
	Bottom ash	1.674			56	26.7	5.8	0.8	0.6	0.1	0.2	2.6	1.3
Raju et al., (2014)	M sand	2.66	12	3.12									
	Bottom ash	2.2	17	2.85									
Kumar et al., (2014)	River sand	2.65	0.72	2.4									
	Bottom ash	2.12	14.1	6.28									
Aswathy and Paul (2015)	Sand	2.59	12	3.787									
	Bottom ash	2.967	18	2.967									

1.1.2 Fresh Concrete Properties

The coal bottom ash has high water absorption properties, leading to lower workability and water loss from bleeding for concrete that contains coal bottom ash than normal concrete. The free water present in the concrete mix was much lesser as water incorporated in the concrete mix has been adsorbed mainly by the coal bottom ash, reducing workability. Besides, lower in workability, the rough surface and irregular shape of the coal bottom ash causes the increase of inter-particle friction (Singh & Siddique, 2014a; Raju et al., 2014).

Contradicting result was found in Andrade et al. (2009) research. Concrete containing bottom ash has a higher water loss due to bleeding than control concrete. This observation can be explained by the high w/c ratio of the concrete mix, 0.76–1.26, which increases with the bottom ash replacement percentage. Another possible cause of high bleeding was the low fineness of bottom ash (Matahula & Olubajo, 2018). However, the fineness modulus of materials used was not reported. Another observation was the increase of initial and final setting time for bottom ash concrete due to the high w/c ratio. It can be an advantage for ready-mix concrete. Table 2 shows the properties of fresh concrete.

1.2 Density

In terms of density, the higher the content of bottom ash in concrete, the lower the density. It was due to bottom ash having a lower specific gravity that gives lower weight per cubic meter of the concrete as a replacement of sand was by volume (Singh & Siddique, 2014a; Andrade et al., 2009; Singh & Siddique, 2016).

1.3 Compressive Strength

At the early age of 7 days, the compressive strength of bottom ash concrete was lower than control concrete. The cause of lower compressive strength was because the pozzolanic activity had not been triggered due to the absence of the alkali activator. Besides, the high water absorption property of bottom ash absorbs the free water. It releases an air pocket into the cement paste cause higher porosity of concrete and lowering compressive strength. But as curing time increases, the strength differences between bottom ash concrete and control concrete change from negative to positive, indicating that bottom ash concrete's strength gain increases over time. The cause of the delay of strength gain was explained by the delay of pozzolanic activity of bottom ash. The hydration of cement produces alkaline calcium hydroxide that activates the pozzolanic activity of reactive silica in bottom ash and forms calcium silicate and aluminate hydrates. The calcium silicate and aluminate hydrates filled the pores in the concrete, improving compressive strength (Singh & Siddique, 2014a).

Table 2 Properties of fresh concrete

Reference	Sample	Remarks	Mix proportion		Bleeding (%)	Slump (mm)
			C: FA: CA	w/c		
Singh and Siddique (2014a)	Control mix A		1: 1: 2.45	0.45	3.14	70
	A20				2.66	59
	A30				2.77	39
	A40				2.5	30
	A50				1.77	24
	A75				1.73	15
	A100				1.66	10
Singh and Siddique (2016)	A0	River sand	1: 1: 2.45	0.45	0.2017 g/cm^2	
	A20				0.1710 g/cm^2	
	A30				0.1780 g/cm^2	
	A40				0.1667 g/cm^2	
	A50				0.1135 g/cm^2	
	A75				0.1113 g/cm^2	
	A100				0.1100 g/cm^2	
	B0	Quarry sand	1: 1.3: 3	0.5	0.5134 g/cm2	
	B20				0.3649 g/cm^2	
	B30				0.2998 g/cm^2	
	B40				0.2629 g/cm^2	
	B50				0.2396 g/cm^2	
	B75				0.1998 g/cm^2	
	B100				0.1039 g/cm^2	
Andrade et al., (2009)	0% CRT		1: 3: 2.65	0.72	5.55	
	25% CRT3	equivalent volume replacement		0.91	7.91	
	50% CRT3			1.12	7.03	
	75% CRT3			1.26	13.34	
	100% CRT3			1.26	9.37	
	25% CRT4	non-equivalent volume replacement	1: 2.74: 2.65	0.76	8.33	

(continued)

Table 2 (continued)

Reference	Sample	Remarks	Mix proportion		Bleeding (%)	Slump (mm)
			C: FA: CA	w/c		
	50% CRT4		1: 2.48: 2.65	0.81	6.64	
	75% CRT4		1: 2.23: 2.65	0.85	9.11	
	100% CRT4		1: 1.77: 2.65	0.84	6.14	
Raju et al., (2014)	Control mix		1: 1.82: 3.54	0.4		110
	BAC 5					95
	BAC 10					88
	BAC 15					80
	BAC 20					72
	BAC 25					68
	BAC 30					65

The compressive strength of coal bottom ash concrete increased significantly after 28 days due to delayed hydration and pozzolanic activity. Also, coal bottom ash as a replacement of river sand with 0.45 w/c ratios shows improvement if not similar to control concrete in the long term. While bottom ash as a replacement for quarry sand with a w/c ratio of 0.5 shows lower compressive strength even in the long term. The lower w/c ratio positively affects the bottom ash concrete compressive strength (Singh & Siddique, 2016).

The seven-day compressive strength of bottom ash concrete with a replacement percentage of 5–20% was slightly greater than control concrete, while 25–30% had lower compressive strength. At 28 days compressive strength, bottom ash concrete with a replacement percentage of 5–25% was found to improve in strength, but as replacement percentage increases, the compressive strength reduced in a linear trend. The difference in compressive strength between control concrete and bottom ash concrete at 28 days was more prominent than seven days due to a pozzolanic reaction from bottom ash being retarded at an early stage (Raju et al., 2014).

Andrade et al. (2009) conducted two different methods to replace the sand. The first method replaced equivalent volume and adjusted bottom ash volume to the moisture content. This method significantly lowers compressive strength as the bottom ash percentage increases. On the contrary, the second method replaces sand with non-equivalent volume and no adjustment to the bottom ash volume to moisture content. This method changes the mix proportion of concrete, where the amount of overall fine aggregates (sand + bottom ash) was reduced while the cement and gravel ratio was kept constant. This method gives a similar, if not higher compressive strength of bottom ash concrete as compared to control concrete. The reasons for higher compressive strength were the ratio of fine aggregate, reduced with the increase of bottom ash replacement percentage, hence having more cement content per cubic meter. Another reason for lower compressive strength was the higher w/c ratio. The w/c ratio was much lower than the w/c ratio in the first method.

Kumar et al. (2014) found that bottom ash concrete has greater compressive strength at any stage than control concrete. The bottom ash content increases the compressive strength but is reduced after a 40% replacement percentage. This finding was inconsistent with previous research that mentioned lower early strength for bottom ash concrete.

Mohd Sani et al. (2011) carried out an experimental study on replacing fine aggregate in concrete with washed bottom ash by 10–50% by mass. The concrete mix was 1:1.7:2.77 with a w/c ratio of 0.52. The compressive strength first increases until it reaches a peak and decreases again with the increase of bottom ash percentage. The optimum replacement percentage of bottom ash was 30%. The compressive strength of concrete with bottom ash was significantly lower than control concrete by 37.63–49.42% at 28 days of curing. Table 3 shows the concrete density and its compressive strength.

1.4 Split Tensile Strength

The use of bottom ash in concrete improved split tensile strength at all ages, but the improvement was not significant (Singh & Siddique, 2014a). Singh and Siddique (2016) research found two different trends for the two concrete mixes. Concrete mix A has the same mix proportion as Singh & Siddique (2014a), hence a similar trend. On the other hand, concrete mix B showed an opposite trend: the increase of bottom ash content reduces the split tensile strength.

Raju et al. (2014) also found that the split tensile strength of bottom ash concrete reduced with the increase of bottom ash content. However, all replacement percentages of bottom ash from 5 to 30% exceeded the split tensile strength of control concrete.

1.5 Flexural Strength

Similar to the compressive strength, the flexural strength of bottom ash concrete exceeds the flexural strength of control concrete. The bottom ash content increases the flexural strength but falls after a 40% replacement percentage (Kumar et al., 2014).

Table 3 Density and compressive strength of concrete

Reference	Sample	Density (kg/m^3)	Compressive strength (MPa)							
			3 days	7 days	14 days	28 days	56/60 * days	90 days	180 days	365 days
Singh and Siddique (2014a)	Control mix A	2402		28.43		38.18		40.69	45.53	
	A20	2360		28.49		36.48		42.89	45.53	
	A30	2338		26.92		37.04		44.97	45.35	
	A40	2315		27.67		37.80		43.33	48.81	
	A50	2290		25.91		37.30		43.58	46.79	
	A75	2244		24.59		37.42		41.82	47.48	
	A100	2206		24.21		34.97		41.51	45.09	
Singh and Siddique (2016)	A0	2327				38.21		40.82	45.54	48.68
	A20	2286				36.46		42.97	45.59	48.74
	A30	2262				37.14		45.16	46.54	48.86
	A40	2236				37.79		43.38	48.79	51.35
	A50	2205				37.25		43.63	46.81	49.10
	A75	2163				37.48		41.92	47.58	51.38
	A100	2095				34.99		41.61	45.21	48.75
	B0	2396				34.04		42.32	45.56	48.71
	B20	2362				32.17		41.12	45.00	46.95
	B30	2340				32.40		41.10	43.99	46.27
	B40	2296				31.40		39.67	43.56	47.43
	B50	2280				29.61		38.77	45.17	48.71
	B75	2219				27.48		37.34	44.64	47.29

(continued)

Table 3 (continued)

Reference	Sample	Density (kg/m^3)	Compressive strength (MPa)							
			3 days	7 days	14 days	28 days	56/60 * days	90 days	180 days	365 days
	B100	2161				26.86		35.80	42.88	45.54
Andrade et al., (2009)	0% CRT	2238	15.9			28.4		32		
Raju et al., (2014)	25% CRT3	2177	12.5			23.2		25.7		
	50% CRT3	2090	9.9			18		23		
	75% CRT3	1964	6.3			11.5		14.9		
	100% CRT3	1869	4.2			8.6		12.5		
	25% CRT4	2220	19.5			27.2		32.1		
	50% CRT4	2138	17			28.5		35.9		
	75% CRT4	2109	16.1			26.1		32.7		
	100% CRT4	2040	21.2			32.6		38.4		
	Control mix			32.3		44.82				
	BAC 5			33.71		49.91				
	BAC 10			33.5		49.49				
	BAC 15			32.71		47.9				
	BAC 20			33.46		46.49				
	BAC 25			29.96		45.15				
	BAC 30			28.29		43.65				
Kumar et al., (2014)	0% BA			23.56	28.18	30.4	32.87			
	10% BA			26.67	30.98	32.4	35.28			

(continued)

Table 3 (continued)

Reference	Sample	Density (kg/m^3)	Compressive strength (MPa)							
			3 days	7 days	14 days	28 days	56/60 * days	90 days	180 days	365 days
	20% BA			28.12	32.14	33.4	36.7			
	30% BA			31.67	33.52	35.17	38.23			
	40% BA			32.14	34.85	36.2	39.16			
	50% BA			28.34	30.42	32.25	33.42			
Mohd Sani et al., (2011)	Control mix		28.07	34.21		39.52	43.52*			
	M10		17.56	16.52		21.41	23.71*			
	M20		19	20.03		23.78	23.92*			
	M30		18.79	20.81		24.65	27.44*			
	M40		15.83	17.87		19.99	23.44*			
	M50		14.4	17.3		21.2	22.44*			

2 Modulus of Elasticity

The increase in bottom ash content reduced the modulus of elasticity of concrete almost linearly (Singh & Siddique, 2014a; Singh & Siddique, 2016). The modulus of elasticity of bottom ash concrete exceeded the control concrete at a low replacement percentage, 5–15%. Still, as the bottom ash percentage increases, the modulus of elasticity decreases (Raju et al., 2014). Table 4 shows the concrete's split tensile strength, flexural strength, and modulus.

2.1 Durability

2.1.1 Permeable Pores/Water Absorption

In Singh and Siddiques (2014a) research, concrete with a 20% replacement percentage for bottom ash slightly reduced permeable pores and water absorption at seven days only. In general, the permeable pores of concrete increase with the increase of bottom ash concrete compared to control concrete. Consequently, the improved connectivity between capillary pores in the cement paste increased the water absorption of concrete. In Raju et al. (2014) findings, the 5% replacement percentage for bottom ash can slightly reduce the water absorption of concrete. Still, as the replacement percentage increased to 30%, the water absorption increased by 23.9%.

As the concrete ages, permeability and water absorption were reduced for control concrete and 40–100% bottom ash concrete. The high bottom ash content shows more prominent changes in permeable pores and water absorption over time. At an early stage, bottom ash absorbs water into the particles and releases air pores into the cement matrix. The high-water absorption capacity and pozzolanic activity were absent. Over time, the pozzolanic activity of bottom ash was activated by the alkaline calcium hydroxide, which is the cement hydration product. It forms calcium silicate and aluminate hydrates, filling the cement matrix's pores produced a more compacted concrete (Singh & Siddique, 2014a).

2.1.2 Abrasion Resistance

The abrasion depth of concrete increases with the bottom ash percentage increase. The compressive strength explained the difference in abrasion depth. The 28 days compressive strength of 100% bottom ash percentage for concrete mix A and concrete mix B were 34.99 MPa and 26.86 MPa, respectively. The corresponding abrasion depth was 1.14 mm and 1.23 mm, respectively. The high w/c content of concrete mix B reduced the abrasion resistance of concrete. The reduction in abrasion depth from 28 to 365 days increases with the bottom ash percentage. Control mix A reduces

Table 4 Split tensile strength, flexural strength, and elastic modulus of concrete

Reference	Sample	Split tensile strength (MPa)					Flexural strength (MPa)				Elastic Modulus (GPa)			
		7 days	28 days	90 days	180 days	365 days	7 days	14 days	28 days	56 days	28 days	90 days	180 days	365 days
Singh and Siddique (2014a)	Control mix A	2.09	2.67	3.09	3.37						29.18	29.82	30.34	
	A20	2.2	2.72	3.2	3.35						27.65	28.13	29.43	
	A30	2.28	2.94	3.27	3.39						27.23	27.86	29.04	
	A40	2.17	2.89	3.29	3.46						26.5	27.05	28.4	
	A50	2.28	2.97	3.31	3.57						25.95	26.91	28.07	
	A75	2.19	2.92	3.24	3.59						25.05	25.48	26.71	
	A100	2.18	2.78	3.24	3.6						23.14	23.75	25.55	
Singh and Siddique (2016)	A0		2.67	3.09	3.36	3.8					29.18	29.82	30.04	31.95
	A20		2.72	3.20	3.34	3.90					27.65	28.13	29.11	30.71
	A30		2.94	3.27	3.38	3.86					27.22	27.83	28.74	30.06
	A40		2.89	3.28	3.45	3.84					26.50	27.03	28.14	29.78
	A50		2.96	3.31	3.56	3.82					25.93	26.91	27.79	29.50
	A75		2.91	3.23	3.58	3.81					25.05	25.48	26.44	28.48
	A100		2.78	3.23	3.60	3.80					23.13	23.75	25.31	27.49
	B0		2.57	3.02	3.30	4.11					25.39	26.46	27.02	28.84
	B20		2.46	3.14	3.43	3.94					25.36	26.02	26.42	28.47
	B30		2.51	3.09	3.38	3.92					24.96	26.13	26.97	28.28
	B40		2.52	2.91	3.27	3.85					25.05	26.22	26.82	28.36
	B50		2.33	2.97	3.40	4.15					25.05	25.99	26.40	27.50
	B75		2.21	2.88	3.16	3.73					23.34	23.72	24.23	23.56

(continued)

Table 4 (continued)

Reference	Sample	Split tensile strength (MPa)					Flexural strength (MPa)				Elastic Modulus (GPa)			
		7 days	28 days	90 days	180 days	365 days	7 days	14 days	28 days	56 days	28 days	90 days	180 days	365 days
	B100		2.17	2.82	3.23	3.69					21.31	21.46	22.00	22.30
Raju et al., (2014)	Control mix		3.24								34.06			
	BAC 5		3.7								36.88			
	BAC 10		3.63								35.44			
	BAC 15		3.59								35.14			
	BAC 20		3.61								33.24			
	BAC 25		3.43								33.09			
	BAC 30		3.39								31.96			
Kumar et al., (2014)	0% BA						2.2	3.1	3.4	4.27				
	10% BA						3.35	5.11	6.25	7				
	20% BA						5.1	6.95	7.66	8.18				
	30% BA						6.8	8.1	8.7	8.98				
	40% BA						7.94	8.8	9.04	9.24				
	50% BA						4.51	6.89	7.4	7.85				

abrasion depth by 35%, while the 100% bottom ash has a 39.47% reduction. Control mix B and 100% bottom ash replacement reduce 20.9% and 40.65%, respectively. It indicates that bottom ash concrete has better abrasion resistance performance in the long term (Singh & Siddique, 2016). Table 5 shows the concrete's permeable pores, water absorption, and abrasion depth.

2.2 *Microstructure*

Singh and Siddique (2014a; b) observe that the microstructure of control concrete shows dense, compact, and continuous CSH gel and giant crystals of portlandite in the cement paste. As for concrete with 30% bottom ash, the microstructure of cement paste consists of dense, compact, and sheeted CSH gel and small foils and flakes of early C_3S hydration product. On the other hand, 50% to 100% bottom ash content showed microstructure with compact equant CSH particles less than 1 μm, smaller in size than control concrete. All concrete with or without bottom ash was observed with ettringite needles filling the voids. The use of bottom ash produces tiny air pores when bottom ash absorbs water and release air. The higher the bottom ash content, the greater the number and size of air pores.

3 Methodology

3.1 *Materials Preparation*

Research materials such as Fine Aggregate, Coarse Aggregate, Coal Bottom Ash (CBA), Potable Water, and Portland Cement were prepared. Sarawak Energy Berhad (SEB) provided the coal bottom ash (CBA). As indicated in Fig. 1, the bottom ash was sun-dried for many days. Any crumbled particles were crushed by hand with a hammer to the appropriate particle size. More extensive unburned pulverized coal was eliminated by sifting through a 4.75 mm sieve.

The fine aggregate for the experiment was river sand, which has a particle size that passes through a 4.75 mm screen. Gravel with particle sizes ranging from 5 to 20 mm were chosen for this experiment as coarse aggregate. 20, 14, 12.5, 10, 6.3, and 5 mm of gravel were sieved. It is to ensure that all concrete mixtures produce uniform results. River sand and gravel were obtained from local suppliers through a hardware shop. Cahya Mata Sarawak (CMS) Berhad provided Portland cement, used as a binder to manufacture concrete. Potable water was utilized during the mixing procedure to guarantee a suitable chemical reaction between all components.

Table 5 Permeable pores, water absorption, and abrasion depth of concrete

Reference	Sample	Permeable pores (%)		Water absorption (%)			Abrasion depth (mm)			
		7 days	180 days	7 days	28 days	180 days	28 days	90 days	180 days	365 days
Singh and Siddique (2014a)	Control mix A	14.25	14.05	5.9		5.82				
	A20	13.52	15.6	5.8		6.01				
	A30	16.23	16.05	6.15		6.17				
	A40	18.41	17.44	6.49		6.29				
	A50	19.16	17.84	6.9		6.3				
	A75	19.7	17.45	7.06		6.3				
	A100	20.02	18.06	7.21		6.46				
Singh and Siddique (2016)	A0						0.80	0.77	0.55	0.52
	A20						0.88	0.79	0.61	0.54
	A30						0.96	0.84	0.64	0.58
	A40						0.99	0.84	0.72	0.60
	A50						1.02	0.89	0.78	0.63
	A75						1.05	0.95	0.86	0.67
	A100						1.14	1.01	0.91	0.69
	B0						0.67	0.65	0.57	0.53
	B20						0.72	0.71	0.62	0.55
	B30						0.75	0.71	0.65	0.56
	B40						0.86	0.79	0.68	0.59
	B50						0.93	0.81	0.66	0.58
	B75						1.01	0.90	0.78	0.65

(continued)

Table 5 (continued)

Reference	Sample	Permeable pores (%)		Water absorption (%)			Abrasion depth (mm)			
		7 days	180 days	7 days	28 days	180 days	28 days	90 days	180 days	365 days
	B100						1.23	0.93	0.82	0.73
Raju et al., (2014)	Control mix				5.9					
	BAC 5				5.82					
	BAC 10				6.15					
	BAC 15				6.49					
	BAC 20				6.9					
	BAC 25				7.06					
	BAC 30				7.21					

Fig. 1 Coal bottom ash

3.2 Sample Fabrication

First, fine aggregate, coarse aggregate, bottom ash, and Portland cement were thoroughly combined in a mixing tray to make CBA concrete. For the mixing procedure, potable water was added to the concoction. The concrete's workability (slump test) was performed before casting into the 100 mm × 100 mm × 100 mm cube molds, the concrete's workability (slump test) was performed. Before demolding, the molded concrete was left for 24 h. The de-molded concrete cubes were placed in the water curing tank for the curing procedure. Finally, compressive strength tests were done 7 and 28 days after the curing process. The IS 10262 (2009) standard was followed in producing the concrete mix. Figure 2 depicts the experimental methods.

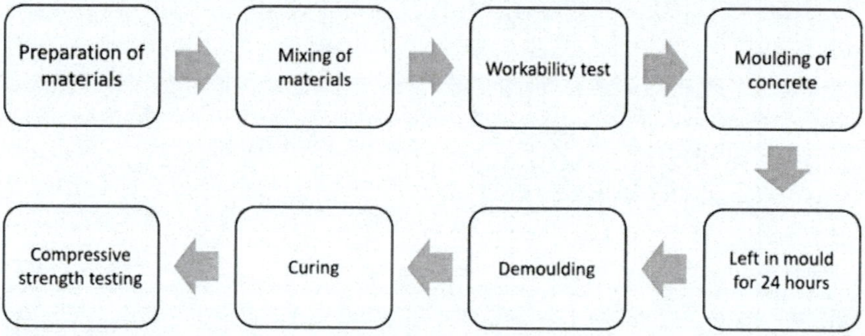

Fig. 2 Experiment procedure

Table 6 Mix proportions for the concrete mixture

Mix	Bottom ash (%)	Cement (kg/m^3)	Fine aggregate (kg/m^3)	Bottom ash (kg/m^3)	Coarse aggregate (kg/m^3)	Water, (kg/m^3)
M0	0	409.2	683.1	0	1103.3	204.6
M1	10	409.2	616	68.2	1103.3	204.6
M2	20	409.2	546.7	136.4	1103.3	204.6
M3	30	409.2	478.5	205.7	1103.3	204.6
M4	40	409.2	410.3	273.9	1103.3	204.6
M5	50	409.2	342.1	342.1	1103.3	204.6

3.3 Nominal Proportions

IS 10262 (2009) and IS 456 (2000) standards created the concrete mix. For the concrete, M30 was used as the grade. With a water-cement ratio of 0.50, the nominal proportions were 1:1.67:2.70 (cement: fine aggregate: coarse aggregate) by mass.

Table 6 shows the proportions of CBA concrete and conventional concrete used in the experiment. In this study, six different concrete mixtures were poured. These mixes are made up of concrete that has been replaced with sand aggregate to the extent of 0, 10, 20, 30, 40, and 50% of the bottom ash. For comparison, concrete with a replacement percentage of 0% was used as ordinary concrete. It eliminated entrapped air and prevented particle segregation during casting, whereas each specimen was compacted using a concrete poker vibrator. Before the compressive strength test, the concrete was cured for seven days and 28 days. In the next part, the findings of the workability and compressive strength tests were obtained and discussed.

3.4 Workability Test

A slump test was carried out on the fresh concrete to ensure the concrete mix workable for casting. According to ASTM C142/C142M-20 (2020), a slump test was carried out.

3.5 Compressive Test

The compressive strength of specimen cubes was determined using Unit Test Autocon 2000. The loading was applied at the rate of 2.4 $\pm$ 0.2 kN/s at 100 mm $\times$ 100 mm area of the specimen. The compressive strength test was carried out according to BS EN 12390-3 (2019). The test results were average and reported for each concrete mix.

4 Results and Discussions

4.1 Workability Test

The slump cone test was used to assess the workability or consistency of this experiment. The test was conducted batch after batch to confirm that the concrete mixture was consistent. The concrete mixture's consistency was determined. The workability of the mixture deteriorated as the replacement of the percentage of bottom ash rose, as seen in Fig. 3. The maximum slump value was found in the standard concrete (M0) combination, 60 mm, as shown in Fig. 4, indicating medium workability.

A steady decrease was seen in the CBA concrete mixture (M1, M2, M3, M4, and M5). The slump value was reported in the 52–59 mm range. As demonstrated in Fig. 5, the concrete M5 with a 50% replacement percentage of bottom ash has the lowest slump value of 52 mm.

It is most likely owing to the CBA's high-water absorption. As a result, during the mixing process, the porous particles of CBA absorb more water than the sand aggregate particles. The rough texture and complex form of CBA particles also increased inter-particle friction (Singh and Siddique, 2014a), resulting in a fall in slump value.

From previous literature, it was known that the CBA was lower in density or specific gravity than sand (Singh & Siddique, 2014a, b; Singh & Siddique, 2016; Andrade et al., 2009; Raju et al., 2014; Kumar et al., 2014; Mahajan & Bahagat, 2020). So, the volume of CBA was larger than sand at the same mass. In the current experiment, CBA replaced sand by mass, which can be deduced that the total volume

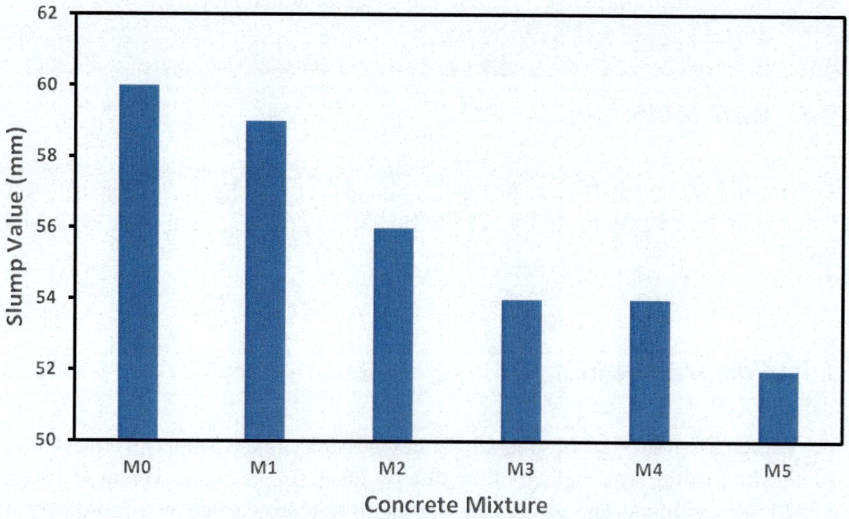

Fig. 3 Graph of slump value with different batches

Fig. 4 Workability of ordinary concrete (M0)

Fig. 5 Workability of CBA concrete (M5)

of fine aggregate was getting more prominent as the replacement by CBA increased. It increases the total surface area for water absorption, thus further reducing the workability. A similar trend was observed in a mortar with CBA replacement (Mahajan & Bahagat, 2020).

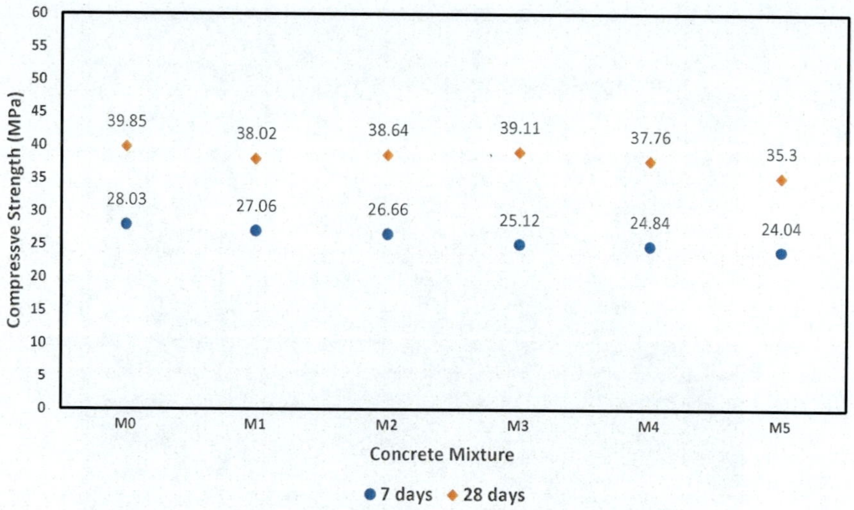

Fig. 6 Graph of compressive strength with replacement percentage of CBA

4.2 *Compressive Strength*

The compressive strength test assessed the concrete's capacity to withstand axial loading. The compressive strength of concrete was determined at seven days and 28 days of curing.

Concrete mixtures with and without CBA have their compressive strength tested. Figure 6 shows the results. At seven days of curing age, concrete mixtures M1, M2, M3, M4, and M5 had compressive strengths that were 3.46, 4.89, 10.38, 11.38, and 14.23% lower than regular concrete mixture M0. The increase in CBA replacement percentage leads to the drop of compressive strength, comparable to the findings of Kajal et al. (2017).

Concrete containing CBA has attained a compressive strength of the targeted concrete grade after 28 days of curing. The reduction of compressive strengths of concrete mixtures M1, M2, M3, M4, and M5 was 4.59%, 3.04%, 1.86%, 5.24%, and 11.42%, respectively, as compared to M0's compressive strength of 39.85 MPa. M3 has the highest compressive strength of all the CBA concrete mixtures, at 39.11 MPa, despite not exceeding M0. This finding was similar to the findings reported by Singh and Siddique (2016), where the highest compressive strength for CBA concrete was 30% CBA replacement for concrete designated strength of 34 MPa and 40% CBA replacement for concrete designated strength of 38 MPa achieved the minor differences in compressive strength as compared to the respective control concrete. Other researchers also have similar findings (Singh & Siddique, 2014a, b; Kumar et al., 2014); hence, the optimum replacement percentage for CBA was 30% to 40%. A notable remark was that the previous literature's CBA replacement was either volume or mass.

The strength development from 7 to 28 days for M0, M1, M2, M3, M4 and M5 were 42.17%, 40.5%, 44.94%, 55.69%, 52.01% and 46.84%, respectively. It is observed that CBA concrete has greater strength development than conventional concrete (M0) except M1 that contains the lowest amount of CBA. The delayed pozzolanic activity of CBA results in a slower strength development at an early stage and accelerates after 28 days.

5 Conclusion and Summary

The study effort provided a better knowledge of the characteristics of normal weight concrete containing bottom ash as a partial replacement for sand aggregate. The experiment was conducted on a concrete mixture with various CBA replacement percentages. The workability of CBA concrete deteriorated as the proportion of CBA replaced increased. It is owing to CBA's high-water absorption and its rough texture and complex form. The compressive strength of the concrete mixture containing CBA dropped somewhat after seven days. On the other hand, the growth of compressive strength of concrete mixtures, including CBA, was unaffected, and the outcome was equal to that of concrete mixtures containing no CBA.

In comparison to concrete without CBA, concrete with CBA took longer to gain strength. Concrete with a 30% partial replacement percentage of CBA had a compressive strength of 39.11 MPa after 28 days, equivalent to concrete without CBA, which had a compressive strength of 39.85 MPa. The use of CBA as an acceptable aggregate substitute helps alleviate the problem of agricultural or industrial waste disposal while also preserving natural resources.

Acknowledgements The authors would like to thank Swinburne University of Technology Sarawak Campus and Universiti Malaysia Sarawak (UNIMAS) for the collaboration.

References

Abubakar, A. U., & Baharudin, K. S. (2012). Potential use of Malaysian thermal power plants coal bottom ash in construction. *International Journal of Sustainable Construction Engineering & Technology, 3*(2), 25–37.

Andrade, L. B., Rocha, J. C., & Cheriaf, M. (2009). Influence of coal bottom ash as fine aggregate on fresh properties of concrete. *Construction and Building Materials, 23*(2), 609–614. https://doi.org/10.1016/j.conbuildmat.2008.05.003

Aswathy, P.U., & Paul, M. M. (2015). Behaviour of Self Compacting Concrete by Partial Replacement of Fine Aggregate with Coal Bottom Ash. *International Journal of Innovative Research in Advanced Engineering (IJIRAE), 2*(10), 2349–2163. Retrieved from www.ijirae.com

IS 10262. (2009). *Indian concrete mix design guidelines*. New Delhi: Indian Standard

IS 456. (2000). *Plain and reinforced concrete—Code of Practice*. New Delhi: Indian Standard

Kajal, S., Vedpal, E., Kumar, E.R. (2017). Strength performance of concrete using bottom ash as fine aggregate. *International Journal of Latest Research in Science and Technology, 6*(5), 17–20. https://www.mnkjournals.com/journal/ijlrst/pdf/Volume_6_5_2017/10753.pdf

Kim, H. K., & Lee, H. K. (2011). Use of power plant bottom ash as fine and coarse aggregates in high-strength concrete. *Construction and Building Materials, 25*(2), 1115–1122. https://doi.org/10.1016/j.conbuildmat.2010.06.065

Kumar, D., Gupta, A., & Ram, S. (2014). Uses of Bottom ash in the Replacement of fine aggregate for Making Concrete. *International Journal of Current Engineering and Technology, 4*(6), 3891–3895.

Mahajan, L.S., Bhagat, S.R. (2020). The contribution of bottom ash toward filler effect with respect to mortar. In: Kanwar, V., Shukla, S. (eds.) *Sustainable civil engineering practices. Lecture notes in civil engineering* (Vol. 72). Singapore: Springer. https://doi.org/10.1007/978-981-15-3677-9_15

Matahula, W., & Olubajo, O. (2018). Effects of limestone and coal bottom ash on setting time of blended portland cement (ternary cement). *Journal of Material Science & Engineering, 7*(5). https://doi.org/10.4172/2169-0022.1000484

Mohd Sani, M. S. H., Muftah, F., & Muda, Z. (2011). The properties of special concrete using washed bottom ash (wba) as partial sand replacement. *International Journal of Sustainable Construction Engineering and Technology, 1*(2), 65–76. https://publisher.uthm.edu.my/ojs/index.php/IJSCET/article/view/64

Rafieizonooz, M., Mirza, J., Salim, M. R., Hussin, M. W., & Khankhaje, E. (2016). Investigation of coal bottom ash and fly ash in concrete as replacement for sand and cement. *Construction and Building Materials, 116*(1), 15–24. https://doi.org/10.1016/j.conbuildmat.2016.04.080

Raju, R., Paul, M. M., & Aboobacker, K. (2014). Strength performance of concrete using bottom ash as fine aggregate. *International Journal of Research in Engineering and Technology, 2*(9), 111–122.

Singh, M., & Siddique, R. (2013). Effect of coal bottom ash as partial replacement of sand on properties of concrete. *Resources, Conversation and Recycling, 72*(1), 20–32. https://doi.org/10.1016/j.resconrec.2012.12.006

Singh, M. (2018). Coal bottom ash. In Waste and supplementary cementitious materials in concrete: Characterisation, properties and applications (pp. 3–50). https://doi.org/10.1016/B978-0-08-102156-9.00001-8

Singh, M., & Siddique, R. (2014). Strength properties and micro-structural properties of concrete containing coal bottom ash as partial replacement of fine aggregate. *Construction and Building Materials, 50*(1), 246–256. https://doi.org/10.1016/j.conbuildmat.2013.09.026

Singh, M., & Siddique, R. (2014). Compressive strength, drying shrinkage and chemical resistance of concrete incorporating coal bottom ash as partial or total replacement of sand. *Construction and Building Materials, 68*(1), 39–48. https://doi.org/10.1016/j.conbuildmat.2014.06.034

Singh, M., & Siddique, R. (2015). Properties of concrete containing high volumes of coal bottom ash as fine aggregate. *Journal of Cleaner Production, 91*(1), 269–278. https://doi.org/10.1016/j.jclepro.2014.12.026

Singh, M., & Siddique, R. (2016). Effect of coal bottom ash as partial replacement of sand on workability and strength properties of concrete. *Journal of Cleaner Production, 112*(1), 620–630. https://doi.org/10.1016/j.jclepro.2015.08.001

Bottom and Fly Ash as Sand and Portland Cement Filler Replacement in High Volume Concrete

Chin Mei Yun, Md Rezaur Rahman, Kuok King Kuok, Amelia Chai Pei Sze, Chong Shi Qin, and Muhammad Khusairy Bin Bakri

Abstract The effects of coal bottom ash (CBA) as a partial sand replacement and coal fly ash (CFA) as a partial Portland Cement replacement in high-volume fly ash (HVFA) concrete are investigated in this paper. CBA and CFA are waste products from coal-fired power plants worth studying because of the large amount produced each year. Furthermore, this study promotes the use of CBA and CFA in the construction industry to recycle and reuse waste products while reducing the use of river sand, which is becoming depleted, and reducing pollution from Portland cement manufacturing. Because its characteristics are compared to river sand, CBA may be used as a sand substitute in concrete, but CFA can harden concrete during the hydration process. The workability and compressive strength of Fly Ash-Bottom Ash (FABA) Concrete are among the experimental efforts. River sand is replaced with CBA in 10, 20, 30, and 40%, whereas Portland Cement is replaced with CFA 40% of the time. The concrete samples were left to cure for 7, 14, 28, 56, and 91 days. To lower the water-to-cementitious material ratio, high-range water reducers, also known as superplasticizers, are recommended in the FABA concrete samples. The numerical analysis helps to determine the average compressive strength and the optimum CBA replacement percentage by using the results of the failure load from the experiment works. The addition of the superplasticizer to the concrete samples had shown better results in compressive strength and workability.

Keywords Coal · Fly ash · Bottom ash · Concrete

C. M. Yun · K. K. Kuok · A. C. P. Sze · C. S. Qin
Faculty of Engineering, Computing and Science, Swinburne University of Technology Sarawak Campus, Jalan Simpang Tiga, 93400 Kuching, Sarawak, Malaysia

M. R. Rahman (✉) · M. K. B. Bakri
Faculty of Engineering, Universiti Malaysia Sarawak, Jalan Datuk Mohammad Musa, 94300 Kota Samarahan, Sarawak, Malaysia
e-mail: rmrezaur@unimas.my

M. R. Rahman et al. (eds.), *Waste Materials in Advanced Sustainable Concrete*, Engineering Materials, https://doi.org/10.1007/978-3-030-98812-8_7

1 Introduction

One of Malaysia's primary sources of electricity is a coal-fired power plant. The Tanjung Bin Power Plant in Johor burns more than 18,000 tons of coal each day. This power station generates around 1620 tons of fly ash (CFA) and 180 tons of bottom ash (CBA) each day. CBA accounts for 10–20% of total ash produced during combustion, whereas CFA accounts for the remaining 80% to 90% of total ash (Singh & Siddique, 2016). The waste product from coal-fired power plants is increasing year after year, and these coal ashes are generally disposed of in an ash pond, polluting thousands of acres of land and causing harm to human health. As a result of these issues and the growing environmental concern, efforts to employ coal ash in the building industry have been concentrated and begun CBA particles are angular, porous, and tend to interlock. CBA has a higher look, and particles amount to smaller than 75 m than river sand. CBA is selected as a substitute for river sand in concrete because of these properties. During the hydration process, CFA hardens the concrete and imparts cementitious characteristics, allowing it to be used as a Portland cement substitute material.

Most recent research has concentrated on using CBA as a partial sand replacement in ordinary concrete, but there has been little study on using CBA in high-volume fly ash (HVFA) concrete. As a result, this research evaluates the fly ash-bottom ash (FABA) concrete characteristics. Furthermore, using CBA and CFA in concrete assists to reduce the use of river sand and Portland Cement, as the manufacturing of Portland Cement emits a significant amount of carbon dioxide into the atmosphere, and river sand is becoming depleted. As a result, this study looks at the feasibility of CBA and CFA as partial replacements for river sand and Portland Cement, respectively, regarding the workability of fresh FABA concrete and compressive strength after curing.

2 Coal Fly Ash (CFA)

The combustion of coal for electric power generation produces coal ash by-products. One of these by-products is coal fly ash. Due to the fine property of fly ash, it escapes the boiler together with flue gas, hence given the name fly ash. Fly ash is collected in an electrostatic precipitator, baghouse or fabric filter (Chou, 2012). The properties of fly ash can vary for each power station. Fly ash is divided into two major classes, Class F and Class C. Class F fly ash is produced from the combustion of anthracite or bituminous coal and possesses pozzolanic property. While Class C fly ash is produced from lignite or sub-bituminous coal combustion, it possesses pozzolanic property and cementitious property (Alonso & Wesche, 1991).

In this study, coal fly ash is chosen as one of the fly ash-bottom ash (FABA) concrete components. CFA has a glassy silicate appearance and pozzolanic characteristics to solidify and create a cementing compound when exposed to moisture

(Kumar et al., 2018). Silicon dioxide (SiO_2), aluminium oxide (Al_2O_3), and calcium oxide (CaO) make up a large portion of the CFA (Rafieizonooz et al., 2016). Class F fly ash is more widely utilised in concrete in the building industry because it provides better overall performance than Class C fly ash and provides higher resistance to sulphate attack.

3 Coal Bottom Ash (CBA)

Coal bottom ash is the coarse granular coal ash collected at the bottom of the furnace. Coal bottom ash is also a by-product of coal combustion. However, the properties of coal bottom ash are very different from coal fly ash. The size of coal bottom ash varies from fine sand to fine gravel (Ghassemi et al., 2004). The quality of coal bottom ash varies, but the coal used for combustion.

The particle size distribution of CBA is comparable to that of fine aggregates such as river sand. CBA's particle size distribution is similar to river sand, implying that CBA has similar particle sizes to river sand. In comparison to river sand, CBA has a glassy texture and lower specific gravity. In terms of microstructural features, CBA has a porosity and angular look, giving it a rough texture and interlocking properties (Singh & Siddique, 2016). The presence of silicon oxide in the CBA is critical in the concrete manufacturing process. As a result, CBA has gotten a lot of attention since it's an excellent partial replacement for river sand in concrete manufacturing.

4 Factor Affecting the Strength of Fly Ash—Bottom Ash (FABA) Concrete

4.1 Percentage of Fly Ash Replacement

Because a large amount of CFA is used as a partial replacement material in FABA concrete, at least 30% CFA must be used as a partial replacement in Portland cement (Thomas, 2007). Both studies have found that increasing the percentage of CFA in concrete from 30 to 50% causes a decrease in compressive strength at an early age compared to OPC concrete (Rashad, 2015; Siddique, 2004).

4.2 Percentage of Bottom Ash Replacement

The amount of CBA used to substitute river sand impacted the developing compressive strength of FABA concrete (Rafieizonooz et al., 2016). After 7 days of curing, FABA concrete with 25% CBA obtained 90.1% of the compressive strength of

OPC concrete, but 50% CBA and 75% CBA only achieved about 82% and 77% of the compressive strength of OPC concrete, respectively. Furthermore, increasing CBA replacement lowered the alkalinity of the pore solution, slowing the hydration process.

4.3 Water to Cementitious Material (W/CM) Ratio

The water-to-cementitious material (w/cm) ratio is essential in the concrete mixture's strength development. The w/cm ratio requirement is generally determined by fly ash replacement in the concrete mixture. To achieve superior strength development in the concrete mixture, the suitable w/cm ratio for high-volume fly ash concrete is proportioned (Thomas, 2007). In HVFA concrete, a lower w/cm ratio is often utilized to achieve superior early-age compressive strength development. According to specific results, the compressive strength of HVFA concrete mixtures increases significantly when the w/cm ratio is reduced (Wang & Park, 2015).

The higher the w/cm ratio, the more porosity and bleeding in the bottom ash concrete mixes (Singh, 2015). Furthermore, because of the presence of CBA in the concrete mixture, there is a good chance that more bleeding water was stored at the bottom of aggregates. Bleeding water may cause additional pores to form on the surface of aggregates, decreasing the bonding action between aggregates and cement paste with concrete mixes. The concrete becomes weaker and porous in this condition, resulting in poor compressive strength development in FABA concrete.

4.4 Curing Period

The curing age may also impact the strength development of FABA concrete. The development of the compressive strength pattern in FABA concrete becomes more pronounced as the curing time increases. The compressive strength of FABA concrete is lower than the control concrete between 7 and 28 days of curing, but after 91 days of curing, FABA concrete began to grow more rapidly than the control concrete (Rafieizonooz et al., 2016). As a result, on the 7th to 28th day of curing, the pozzolanic action of the coal ash concrete is slower than the control concrete. Compared to OPC concrete, the calcium silicate hydrate (C–S–H) gel in FABA concrete is not as compact and monolithic. Furthermore, the presence of CBA increases the number and size of voids, affecting the compressive strength development of FABA concrete in the early curing stage.

4.5 *Addition of Superplasticizers*

High range water reducers, also known as superplasticizers, can be used in concrete to lower the w/cm ratio. Compared to concrete mixes without superplasticizers, the inclusion of superplasticizers increases the compressive strength of bottom ash concrete by 10–20%. (Singh & Siddique, 2016). The use of superplasticizers helps minimize the w/cm ratio while also reducing bleeding, resulting in improved compressive strength development in the concrete mixture.

5 Methodology

5.1 *Materials*

Materials such as river sand as fine aggregate, gravel as coarse aggregate were obtained from local suppliers through the hardware store. Portable water was obtained from the Swinburne University laboratory. Ordinary Portland cement was obtained from Cahya Mata Sarawak (CMS) Berhad. Sarawak Energy Berhad (SEB) provided the coal fly ash (CFA) and coal bottom ash (CBA). The superplasticizer used was Naphthalene Sulfonate Formaldehyde branded as Might Kao 150, manufactured by Kao Industrial (Thailand).

5.2 *Sample Fabrication, Mix Proportion and Curing*

The experimental work investigated the characteristics of the fly ash–bottom ash (FABA) concrete and investigated its compressive strength development at different stages of curing time by manipulating the proportion of coal bottom ash (CBA) as a partial replacement of river sand. There are certain fixed factors in the experiment, such as the water to cementitious material (w/cm) ratio, the amount of coarse and fine aggregates, the aggregate Particle Size Distribution (PSD) curve, and the kind of fly ash and bottom ash.

High-volume fly ash (HVFA40) concrete with 40% fly ash partial Portland cement replacement and 0% CBA replacement is utilized as the control set for comparative reasons. 100 × 100 × 100 mm cubes are employed, and there are five batches of concrete casting with 15 samples each batch. The remainder is FABA concrete with a continuous 40% fly ash partial Portland cement replacement and 10, 20, 30, and 40% CBA as a partial river sand replacement. Aside from that, the concrete samples were water-cured for 7, 14, 28, 56, and 91 days. Three samples from each batch are tested for compressive strength after the water curing process, and the average findings are obtained. Table 1 shows the sample proportion.

Table 1 The proportion of fine aggregates for the experiment

Concrete sample	Sand proportion (%)	Coal Bottom ash proportion (%)
HVFA40	100	0
FABA10	90	10
FABA20	80	20
FABA30	70	30
FABA40	60	40

Furthermore, the second phase of this experimental investigation involves the addition of superplasticizers to FABA concrete to evaluate the impacts on the FABA concrete's early age compressive strength development. This portion of the experimental investigation cast four batches of FABA concrete containing 10, 20, 30, and 40% CBA as a partial river sand replacement. In this stage of the concrete mixing, the superplasticizer was added at a rate of around 2% of the cementitious material's weight. However, because this portion of the experimental investigation is focused on the outcome of early age compressive strength of FABA concrete with superplasticizer, the water curing time was set at 7, 14, and 28 days.

5.3 Workability Test

The slump test determined the workability of fresh concrete. Slump test was performed according to ASTM C142/C142M-20 (2020).

5.4 Compressive Test

A compressive strength test was conducted using Unit Test Autocon 2000 regarding BS EN 12390-3 (2019). During the compressive strength test, the loading increased at 2.4 ± 0.2 kN/s. The results for each batch were averaged and recorded.

6 Results and Discussions

6.1 Workability

The water to binder ratio of the concrete samples is set to 0.5, allowing for a comparison of each sample's workability. The concrete samples' workability test results are presented in Table 2. It was observed that the workability of concrete reduced with

Table 2 Workability of the concrete samples

Concrete	Slump height, mm	
	Without superplasticizer	With superplasticizer
HVFA40	65	–
FABA10	61	75
FABA20	57	71
FABA30	50	60
FABA40	43	52

the increase of CBA percentage in the concrete mixture. CBA was made up of dry and porous particles that absorb water inside (Singh & Siddique, 2016). Hence, CBA has higher water absorption properties than river sand. Besides, the higher specific surface area of CBA lead to the increase of water requirement for mixing Park et al. (2021). Furthermore, the CBA particles were irregular and had a rough surface. These properties improve the interlocking and friction between particles, hindering the flow of fresh concrete (Kim & Lee, 2011).

Superplasticizers help improve workability and enhance the concrete strength by reducing the w/cm ratio while maintaining the desired workability. After the superplasticizer was applied to the FABA concrete samples, the workability of the samples with the same fixed w/cm ratio as the first batch of concrete samples rose noticeably. According to the results, the workability of FABA10, FABA20, FABA30, and FABA40 has increased by 23%, 25%, 20%, and 21%, respectively. The use of superplasticizer causes deflocculation and dispersion of cement particles, leading to improved workability of concrete (Neville, 2011).

6.2 *Compressive Strength*

According to Fig. 1, FABA40 concrete samples exhibited the lowest compressive strength of all the samples at all ages. The replacement of sand by CBA was by mass, and CBA had lower specific gravity than sand. Furthermore, CBA increases the need for mixing water during the casting process, allowing pores to form and increasing the porosity of concrete samples (Andrade et al., 2009). These are resulting in the increased volume of aggregate with the increase of CBA replacement. The compressive strength of concrete was generally lower with CBA replacement.

The HVFA40 concrete samples exhibited a lower early age strength development than FABA10 and FABA20 concrete samples but greater than FABA30 and FABA40 at 7 and 14 days. The higher compressive of FABA10 and FABA20 at an early age could probably improve the overall aggregate grading and the irregular form, an interlocking feature of CBA. The strength development of all concrete samples was slow at the early stage due to the delay in the hydration process caused by the high substitution of Portland cement with CFA.

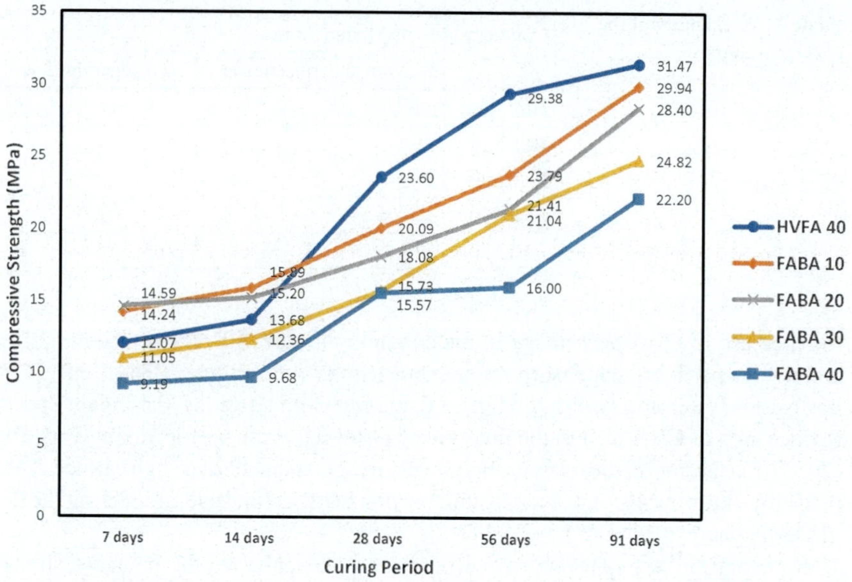

Fig. 1 Average compressive strength of concrete samples in different curing periods

The strength development of all concrete samples from 14 to 91 days was noticeably increased at a greater rate than the strength development from 7 to 14 days. It was contributed to the pozzolanic activity of CFA and CBA (Rafieizonooz et al., 2016). The strength development from 56 to 91 days was significantly for the FABA concrete sample, which the slow pozzolanic activity of CBA can explain compared to CFA. According to Singh and Siddique (2016), the pozzolanic activity of CBA started from 90 days, which was observed in the current study. The increase in bottom ash replacement reduces the pore solution's alkalinity, hence delaying the pozzolanic activity (Singh & Siddique, 2016). It was observed that at 91 days, the compressive strength of FABA10, FABA20, FABA30, FABA40 was 4.9%, 9.8%, 21.1%, 29.5% lower than HVAF40. Lower CBA replacement shows slight strength loss, while higher CBA replacement shows much more significant strength loss. FABA10 and FABA20 reached concrete Grade 25, while FABA30 and FABA40 reached concrete Grade 20 at 91 days.

The lower compressive strength of FABA concrete than HVFA concrete was possibly due to the concrete mixture having low cement content. High cement content concrete incorporated with CBA achieved similar compressive strength as concrete without CBA replacement (Ghafoori & Bucholc, 1997). Replacement of sand by CBA can be done by volume instead of mass due to the lower specific gravity of CBA can improve the concrete strength.

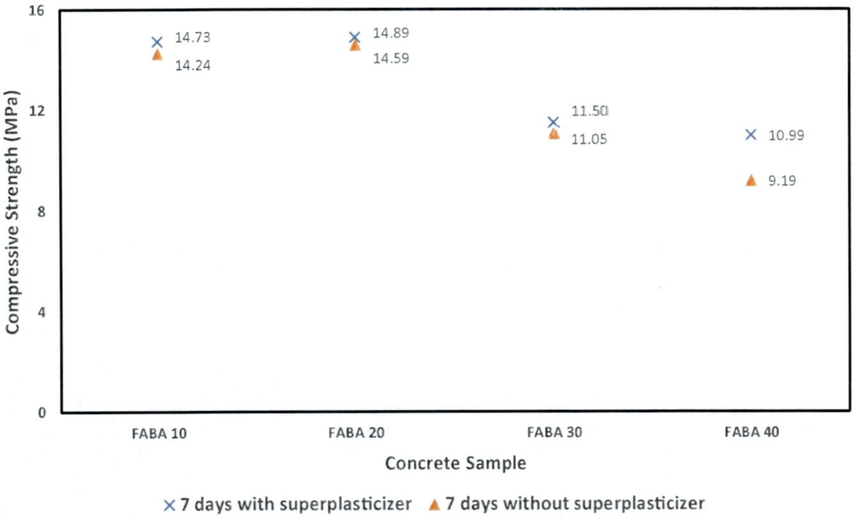

Fig. 2 Average compressive strength of FABA concrete at 7 days

6.3 Addition of Superplasticizer

Figure 2 shows the average compressive strength of the FABA concrete samples with and without superplasticizer. It was observed that the addition of superplasticizer increased slightly in the compressive strength of FABA concrete at 7 days. The improvement in compressive strength at 7 days for FABA10, FABA20, FABA30 and FABA40 were 3.4%, 2.1%, 4.1% and 19.6% respectively. As a result, the superplasticizer had little influence on the 7 day compressive strength development.

However, at 14 days, the superplasticizer has greatly influenced the compressive strength of FABA concrete samples, particularly the FABA40 concrete samples. The improvement in compressive strength at 14 days for FABA10, FABA20, FABA30 and FABA40 were 13.5%, 13.7%, 13.7% and 47.8% respectively. It was observed that FABA40 has greater compressive strength than FABA30 with superplasticizer, while without superplasticizer, FABA40 has lower compressive strength than FABA30 (Fig. 3).

The effect of the superplasticizer was more negligible at the 28 days than it had been on the 14 days. The improvement in compressive strength with the use of superplasticizer at 28 days for FABA10, FABA20, FABA30 and FABA40 were 11.7%, 11.2%, 5.8% and 10.1%, respectively. FABA10 and FABA20 with superplasticizer was 4.9 and 14.8% lower than HVFA40 but obtained 22.44 MPa and 20.11 MPa compressive strength. All three concrete samples are satisfactory for Grade 20 concrete. As for FABA30 and FABA40 with superplasticizer, compressive strength of 16.64 and 17.14 MPa was obtained, respectively, which can be considered for Grade 15 concrete. However, due to the nature of CFA and CBA, a higher concrete strength for FABA with superplasticizer was expected at a much later age (Fig. 4).

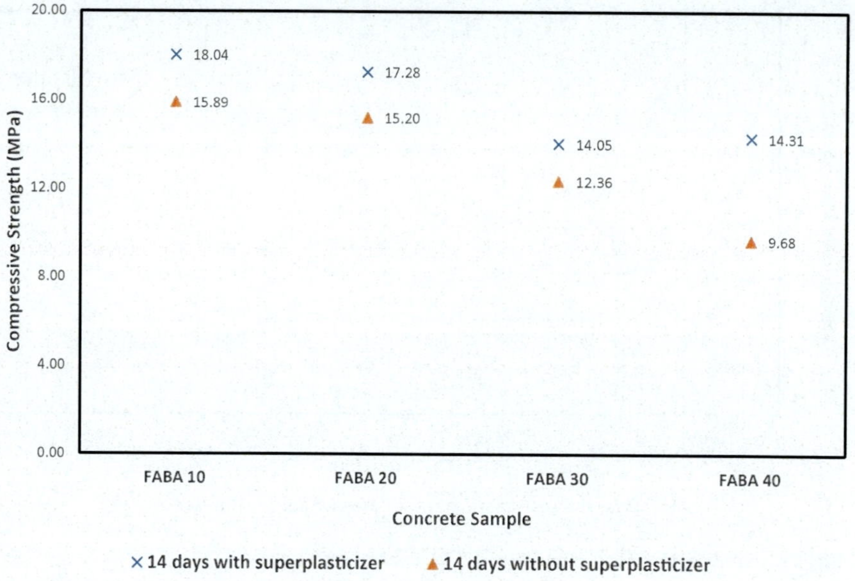

Fig. 3 Average compressive strength of FABA concrete at 14 days

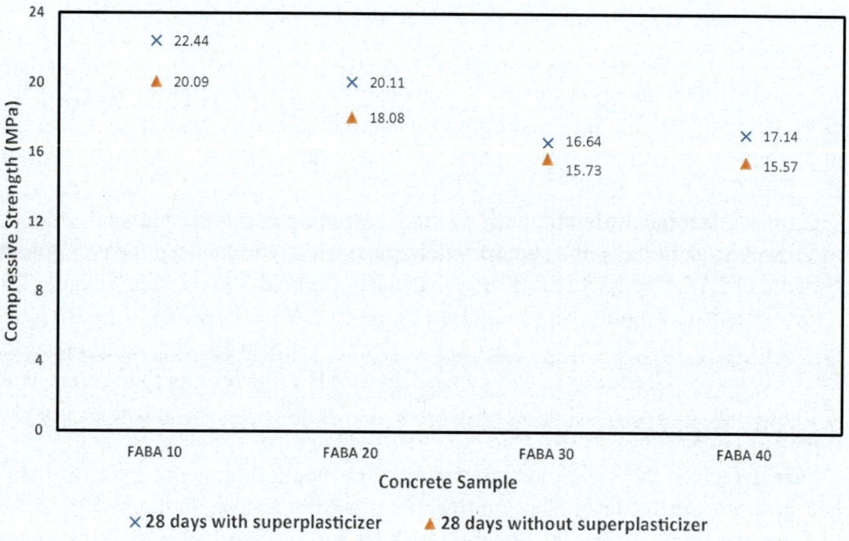

Fig. 4 Average compressive strength of FABA concrete at 28 days

The superplasticizer in FABA concrete demonstrates that it can improve compressive strength. The addition of 2% of superplasticizer slightly increases the overall w/cm ratio, which provides more water for cement hydration. Another possible reason was that superplasticisers induce deflocculation and dispersion of cement particles, allowing more uniform hydration. Lastly, the improved workability aids in the compaction of the concrete matrix during casting; better compaction resulting higher compressive strength (Muhit, 2013).

7 Conclusion

The experimental study on the partial replacement of fine aggregate by CBA in HVFA concrete was conducted. It provided some knowledge about FABA concrete in terms of its workability and compressive strength. When the amount of CBA in the FABA concrete is increased, the workability and compressive strength of the concrete decreases. FABA concrete with 10% and 20% CBA replacement has a more robust early age compressive strength development and can reach Grade 25 concrete at 91 days, acceptable for reinforced concrete work. While FABA concrete with 30% and 40% CBA replacement has a much lower concrete strength, it can still outperform Grade 20 concrete, which is ideal for lighter constructions such as home floors, foundations for interior floor slabs, and roads. At 14 and 28 days, adding the superplasticizer to the FABA concrete improved the compressive strength development.

8 Recommendations and Future Works

Simple experimental works and investigations were carried out in this experimental research project. However, some detailed inquiry or testing could not be carried out due to equipment and machinery constraints. Some recommendations encourage future researchers to conduct experimental work to further their research. For example, it is suggested that an investigation into the internal structure of FABA concrete be conducted to study the behaviour and properties of concrete samples at various curing ages. Magnetic resonance imaging and scanning electron microscopy, for example. Aside from that, research on the compressive strength and characteristics of FABA concrete with more than 40% CBA substitution of river sand is possible. The Rapid Chloride Permeability (RCP) test should also be used to assess the durability of the FABA concrete. This test may be used to assess if concrete can resist chloride ion penetration and, in real-world applications, whether it can protect the reinforcement from chloride assault.

Acknowledgements The authors would like to thank Swinburne University of Technology Sarawak Campus and Universiti Malaysia Sarawak (UNIMAS) for the collaboration.

References

Alonso, J. L., & Wesche, K. (1991). Characterization of fly ash. In K. Wesche (Ed.), Fly ash in concrete: Properties and performance (p. 255). Taylor & Francis.

Andrade, L. B., Rocha, J. C., & Cheriaf, M. (2009). Influence of coal bottom ash as fine aggregate on fresh properties of concrete. *Construction and Building Materials, 32*(2), 609–614. https://doi.org/10.1016/j.conbuildmat.2008.05.003

ASTM C142/C142M-17. (2017). *Standard test method for clay lumps and friable particles in aggregates*. ASTM International, West Conshohocken. https://doi.org/10.1520/C0142_C0142M-17

BS EN 12390-3. (2019). *Testing hardened concrete. Compressive strength of test specimens*. British Standard Institution.

Chou, M.-I.M. (2012). Fly Ash. In R. A. Meyers (Ed.), *Encyclopedia of sustainability science and technology* (pp. 3820–3843). Springer.

Ghafoori, N., & Bucholе, J. (1997). Properties of high-calcium dry bottom ash for structural concrete. *ACI Materials Journal, 94*(2), 90–101. https://doi.org/10.14359/289

Ghassemi, M., Andersen, P. K., Ghassemi, A., & Chianelli, R. R. (2004). Hazardous waste from fossil fuels. *Encyclopedia of Energy*, 119–131. https://doi.org/10.1016/B0-12-176480-X/00395-8

Kim, H. K., & Lee, H. K. (2011). Use of power plant bottom ash as fine and coarse aggregates in high-strength concrete. *Construction and Building Materials, 25*(2), 1115–1122. https://doi.org/10.1016/j.conbuildmat.2010.06.065

Kumar, S., Singh, S. K., & Mishra, S. C. (2018). Processing and characterization of fly-ash compacts. *Materials Today: Proceedings, 5*(2), 3396–3402. https://doi.org/10.1016/j.matpr.2017.11.584

Muhit, I. B. (2013). Dosage limit determination of superplasticizing admixture and effect evaluation on properties of concrete. *International Journal of Scientific and Engineering Research, 4*(3).

Neville, A. M. (2011). *Properties of concrete* (5th edn.). Pearson.

Park, J.-H., Bui, Q.-T., Jung, S.-H., & Yang, I.-H. (2021). Selected strength properties of coal bottom ash (CBA) concrete containing fly ash under different curing and drying conditions. *Materials, 14*(18). https://doi.org/10.3390/ma14185381

Rafieizonooz, M., Mirza, J., Salim, M. R., Hussin, M. W., & Khankhaje, E. (2016). Investigation of coal bottom ash and fly ash in concrete as replacement for sand and cement. *Construction and Building Materials, 116*(1), 15–24. https://doi.org/10.1016/j.conbuildmat.2016.04.080

Rashad, A. M. (2015). A brief on high-volume Class F fly ash as cement replacement – A guide for Civil Engineer. *International Journal of Sustainable Built Environment, 4*(2), 278–306. https://doi.org/10.1016/j.ijsbe.2015.10.002

Siddique, R. (2004). Performance characteristics of high-volume Class F fly ash concrete. *Cement and Concrete Research, 34*(3), 487–593. https://doi.org/10.1016/j.cemconres.2003.09.002

Singh, M. (2015) Effect of coal bottom ash on strength and durability properties of concrete. *PhD Thesis*, Thapar Institute of Engineering and Technology (pp. 1–246). http://hdl.handle.net/10603/230339

Singh, M., & Siddique, R. (2016). Effect of coal bottom ash as partial replacement of sand on workability and strength properties of concrete. *Journal of Cleaner Production, 112*(1), 620–630. https://doi.org/10.1016/j.jclepro.2015.08.001

Thomas, M. (2007) Optimizing the use of fly ash in concrete. *Concrete*. Portland Cement Association.

Wang, X.-Y., & Park, K.-B. (2015). Analysis of compressive strength development of concrete containing high volume fly ash. *Construction and Building Materials, 98*(1), 810–819. https://doi.org/10.1016/j.conbuildmat.2015.08.099

Sawdust as Sand Filler Replacement in Concrete

Chin Mei Yun, Md Rezaur Rahman, Durul Huda, Kuok King Kuok, Amelia Chai Pei Sze, Jong Ka Seng, and Muhammad Khusairy Bin Bakri

Abstract This study investigated sawdust waste as a partial replacement for sand as fine aggregates in producing lightweight sawdust concrete (SC). In Malaysia, sawmills create a considerable amount of wood waste, which has caused environmental issues as it is turned into burning materials. This investigation used sawdust to substitute river sand in the concrete mix proportions of 5%, 10%, 15%, and 20% by weight. The impact of compressive strength, workability, and density was investigated by changing the amount of sand replacement with sawdust. Concrete cubes of 100 mm × 100 mm × 100 mm were produced for compressive strength tests. The compressive strength was measured after 7, 14, and 28 days of cure. Furthermore, the strength and density of conventional concrete and sawdust concrete were examined. The study's findings demonstrated increased sawdust content in concrete decreased compressive strength and density. According to the findings of the experiments, the optimal sawdust content is found to be 5%, with the highest strength of 17.2 MPa after 28 days. As a consequence of the results and observations, sawdust concrete may be used for lightweight structural applications.

Keywords Sawdust waste · Fine aggregates · Concrete · Compressive strength · Density

C. M. Yun · K. K. Kuok · A. C. P. Sze · J. K. Seng
Faculty of Engineering, Computing and Science, Swinburne University of Technology Sarawak Campus, Jalan Simpang Tiga, 93400 Kuching, Sarawak, Malaysia

M. R. Rahman (✉) · M. K. B. Bakri
Faculty of Engineering, Universiti Malaysia Sarawak, Jalan Datuk Mohammad Musa, 94300 Kota Samarahan, Sarawak, Malaysia
e-mail: rmrezaur@unimas.my

D. Huda
Department of Mechanical Engineering and Product Design Engineering, Swinburne University of Technology, Hawthorn, VIC 3122, Australia

M. R. Rahman et al. (eds.), *Waste Materials in Advanced Sustainable Concrete*, Engineering Materials, https://doi.org/10.1007/978-3-030-98812-8_8

1 Introduction

The building sector continues to develop across the world. Concrete is a widely utilized building material in the construction industry. Cement, water, and fine and coarse aggregates are all basic materials used to create concrete (Bedi et al., 2013; Collivignarelli et al., 2020; Priyadharshini et al., 2021; Yalley & Sam, 2018). Indeed, the cost of construction materials is growing as the world's population grows, causing rapid housing and development (Osei & Jackson, 2016). Developers and engineers have expressed worry about this situation, as the depletion of unrenewable natural sources for concrete concerns and climate change (Ahuja & Tatsutani, 2009). It has caused building materials to be more expensive. Therefore, utilising an alternate material readily available locally (Akadiri et al., 2012; Chel & Kaushik, 2018; Danso & Obeng-Ahenkora, 2018; Imbabi et al., 2012).

Palm kernel shell, coconut shell, sawdust, and fly ash are some of the native elements that may be found in industry or agricultural waste (Asuzu et al., 2017; Gana et al., 2018; Ganiron, 2014; Kaniapan et al., 2021; Mannan & Ganapathy, 2002; Reddy et al., 2014; Soliman & Moustafa, 2020). Using waste materials like sawdust in concrete might be a viable solution not just to the pollution problem but also to the high cost of construction materials (Awal et al., 2016; Chowdhury et al., 2015; Mangi et al., 2019; Mwango & Kambole, 2019; Oyedepo et al., 2014; Sojobi & Alavi, 2016; Suliman et al., 2019; Turgut & Algin, 2007; Ugwu, 2019). Sawdust is a waste product that results from the sawmill's cutting, grinding, and sawing of wood (Ahmed, 2021). The wood particles are generally produced from the chip or sawdust form of wood waste, as shown in Fig. 1. Sawdust was undoubtedly used to make charcoal briquettes and cooking fuel (Ganiron, 2014). However, most sawdust was disposed of in the landfills or burned openly (Abdel-Shafy & Mansour, 2018). As a result, assessment is needed to ensure the feasibility of using sawdust waste as

Fig. 1 Wood sawdust

lightweight concrete for building applications (Ahmed et al., 2018; Marveas, 2020; Udoeyo & Dashibil, 2002).

Sawdust can be used as a fine aggregate substitute in concrete as an alternative waste (Prusty et al., 2016). Due to a significant bark volume, the cement hydration may be affected. Thus, it is important to wash the sawdust before putting it into concrete (Udoeyo & Dashibil, 2002). Furthermore, sawdust concrete is light and provides enough heat insulation but low strength (Ganiron, 2014). Boob (2014) investigated adding sawdust waste to concrete on its characteristics. Concrete's compressive strength, water absorption capacity, and density were measured. The findings for sawdust concrete density indicated that increasing the amount of sawdust in concrete causes density to decrease (Boob, 2014).

Gopinath et al. (2015) looked at using dry sawdust as a partial substitute for fine aggregates at 10%, 30%, and 50%. A total of 48 cubes were created. The compressive test revealed that the strength of sawdust concrete dropped as the proportion of sawdust increased (Gopinath et al., 2015). Sawdust concrete is also advised for structural components such as beams and columns (Gopinath et al., 2015). Ambiga and Meenakshi (2015) conducted a study to find a suitable solution of concrete optimum strength. Therefore, they compare the normal and sawdust concrete by using sawdust as a fine aggregate's replacement. At 7 and 28 days, cubes' weight and compressive strength were investigated. As a consequence of the findings, it appears that adding sawdust to concrete reduces the weight of the concrete. In addition, the strength of sawdust concrete has grown with age but has reduced with the addition of sawdust (Ambiga & Meenakshi, 2015).

Abdullahi et al. (2013) investigated and developed sawdust instead of sand in concrete manufacturing. Compared to regular concrete, sawdust inclusion ranges from 0 to 50% in 10% increments. The strength result showed that the strength is reduced when a high proportion of sawdust replaces sand in concrete. Meanwhile, with a strength of 7.41 N/mm^2, the best substitute was determined to be 10% (Abdullahi et al., 2013). As a result, the goal of this experimental investigation was clearly defined. It is critical to determine the proportion of sawdust used to substitute sand in concrete for this investigation.

Furthermore, the optimal proportion of sawdust substitution for evaluating the concrete compressive strength may be discovered. This experiment aims to make sure that sawdust concrete has the same strength as regular concrete. Another objective of this research is to look at the behaviour of concrete when fine aggregate is replaced with sawdust in percentages of 5%, 10%, 15%, and 20%. Furthermore, this study aims to determine the optimal level of sawdust inclusion in concrete without compromising its strength.

2 Methodology

2.1 Material

Gravel as Coarse aggregates, River sand as fine aggregates, Ordinary Portland Cement (OPC), and sawdust were the raw materials utilized in this experiment. Coarse aggregate, fine aggregate, and Portland cement were obtained from Cahya Mata Sarawak (CMS) Berhad. Gravel, ranging in size from 14 to 20 mm, was used as the coarse aggregate for this investigation. The sawdust waste in powder form was collected from a local sawmill. The sawdust waste was later sun-dried and was kept in a sealed bag. Whist, river sand as fine aggregate was used according to IS 383 (2003) grading zone. Fine aggregate, which consists of river sand and sawdust debris were sieved, and particle sizes used were between 4.75 mm and 0.75 um in size., which in detail it was sieved to retain the particle size of 4.75 mm, 2.36 mm, 1.18 mm, 0.6 mm, 0.3 mm, 0.15 mm, 0.075 mm. The particle size distribution curve of river sand and sawdust is shown in Fig. 2. As shown in Fig. 2, sawdust may be used instead of sand since the fineness of both materials is similar. The findings from the sieve size study of sawdust mixed sand are given in Table 1.

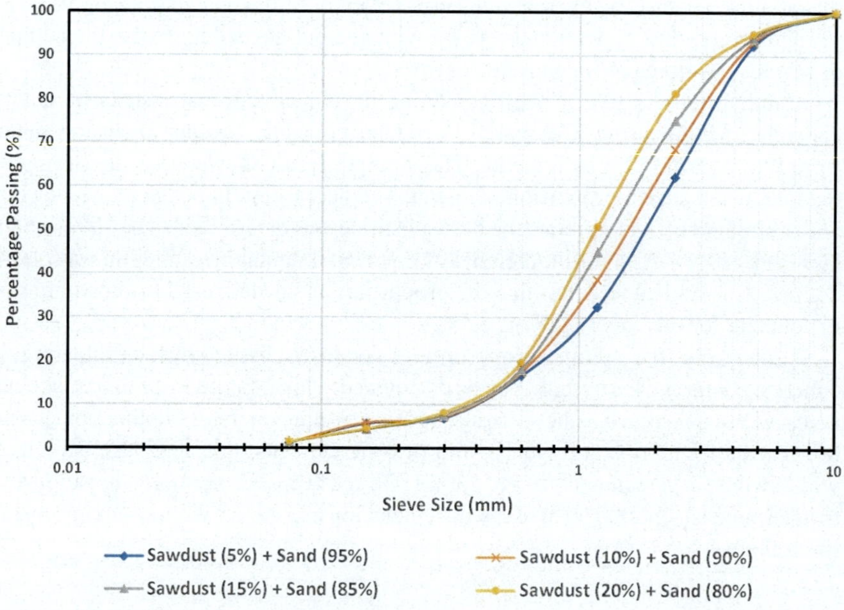

Fig. 2 Particle size distribution of fine aggregates

Table 1 Sieve size analysis of sawdust mixed sand

Sieve sizes	Sawdust (5%) + sand (95%)	Sawdust (10%) + sand (90%)	Sawdust (15%) + sand (85%)	Sawdust (20%) + sand (80%)
10	100	100	100	100
4.75	92.41	93.3	94.14	95.1
2.36	62.33	68.74	75.14	81.55
1.18	32.5	38.77	45.06	50.97
0.6	16.74	17.51	18.2	18.79
0.3	6.9	7.23	7.59	8.31
0.15	5.66	5.89	4.31	4.45
0.075	1.23	1.33	1.46	1.53

2.2 Mix Proportion, Mixing and Curing

Ordinary Portland cement (OPC) concrete, standard according to IS 10262 (2009) and IS 456 (2000), was cast and tested in this experiment, which targets the M20 concrete design grade. The concrete was made using a weight ratio of 1: 1.8: 3 and a water/cement ratio of 0.55. Furthermore, the proportion of sawdust used to substitute sand was changed from 0 to 5%, 10% to 10%, 15% to 20%. Compressive strength was evaluated on a total of 45 specimens. The OPC, fine aggregate and coarse aggregate were weighted proportions for mixing, as stated in Table 1. The electric concrete mixer was used to mix the materials during the casting process. Subsequently, portable water was added to the concrete mixture. The concrete mixture was poured into a 100 mm × 100 mm × 100 mm iron mould. The mould was set up filled the inner surface with oil for molding, to prevent it from sticking. The poker concrete vibrator was used to conduct concrete compaction and remove excess bubbles from the concrete mixture. The concrete was demoulded and placed in a curing tank with total water for 7 days, 14 days, and 28 days after 24 h for curing. The compressive strength of cured concrete was evaluated after 7 days, 14 days, and 28 days.

2.3 Workability (Slump) Test

Before casting, the slump test was performed after the mixing procedure to ensure that each concrete batch was workable. ASTM C142/C142M-20 (2020) standard was used as a reference.

2.4 Compressive Strength Test

A compression strength test was performed to determine the sawdust filler replacement in concrete compression strength. The compressive strength tests were performed using Unit Test Autocon 2000 under BS EN 12390-3 (2019). During the compressive strength test, the load was increased progressively at 2.4 ± 0.2 kN/s across an area of 100 mm × 100 mm until the samples failed. The cube samples were loaded to failure on the 7th, 14th, and 28th days in compressive strength. All cube samples' compressive was measured, and the results were presented.

2.5 Density

The hardened concrete densities were determined according to BS EN 12390-7 (2009). The mass and volume of cube samples cured for 7 days, 14, and 28 days were measured and recorded. The densities of concrete were also determined and presented.

2.6 Scanning Electron Microscopy (SEM) and Energy Dispersive X-Ray/Spectroscopy (EDS/EDX)

SEM and EDX/EDS were carried out in compliance with ASTM C1723-16 (2016) and ASTM E1508-12 (2019) standards, respectively. 100×, 250×, 500×, and 10,000× magnifications were used to analyze the samples. Automated software scans and analyses the elemental composition percentages of the samples. The EDX/EDS was conducted on each sample many times at various phases, with the most representative results chosen. Hitachi TM4000Plus Tabletop Microscope (Hitachi Ltd., Tokyo, Japan) with a Quantax75TM Series Energy Dispersive X-Ray Spectrometer was used in the SEM and EDX/EDS investigations of the materials.

2.7 Fourier Transform Infrared Spectroscopy

Fourier-transform infrared spectroscopy (IRAffinity-1, Shimadzu Corporation, Kyoto, Japan) was used for the FTIR analysis of the samples. Fourier-transform infrared spectroscopy was conducted according to the ASTM E168-16 (2016) and ASTM E1252-98 (2013) standards for qualitative and quantitative analysis. The spectrum scanning was conducted in the wavenumber range of 4000–400 cm^{-1} for each sample. Fourier-transform infrared spectroscopy utilized the samples' infrared spectrum transmittance and absorption to develop a unique molecular fingerprint spectrum. The test was repeated numerous times for each sample, and the most representative results were selected.

3 Results and Discussions

3.1 Workability

The slump test is used to determine the workability of new concrete. The height of the sag between the top of the cone and the top of the concrete was measured. Table 2 provided the data on slump height with various sawdust replace sand in concrete.

As the amount of sand substituted by sawdust was increased, the slump value of fresh concrete reduced, as shown in Fig. 3. The first batch of OPC concrete was categorized as high workability with a slump value of 99 mm based on the data gathered. The following slump value is reduced by 5% of sawdust inclusion from 5 to 20%. Because sawdust has a high-water absorption rate, it dehydrates the concrete additive, the lowering trend of slump value was fair. Medium workability

Table 2 Slump value on various sand replacements by sawdust

Sand replacement by sawdust (%)	Slump value (mm)
0 (0.55)	99
5 (0.55)	80
10 (0.55)	64
15 (0.55)	50
20 (0.55)	40

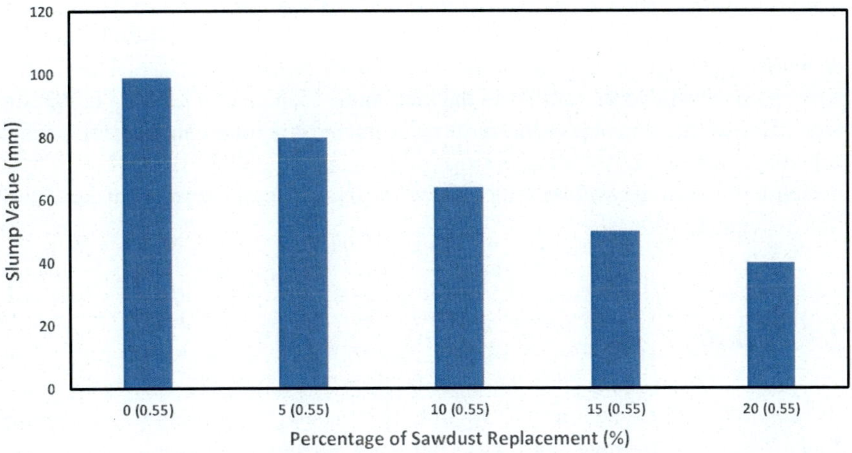

Fig. 3 Slump value with different percentages of sawdust as a replacement

Table 3 Compressive strength on various proportions of sawdust replacement

Batch	Percentage of sawdust (%)	Percentage of sand (%)	Compressive strength test (MPa)		
			7 days	14 days	28 days
SC 0	0	100	26.00	28.40	32.79
SC 1	5	95	10.45	13.36	16.18
SC 2	10	90	5.04	6.58	9.04
SC 3	15	85	2.15	2.10	2.39
SC 4	20	80	2.10	2.13	2.17

was assigned to the concrete with 5%, 10%, and 15% sawdust content, respectively. The slump value achieved was 40 mm when the sawdust content reached 20%. Low workability was characterized as a range of slump concrete. In this scenario, the cube specimen's strength may be harmed by impractical workability.

3.2 *Compressive Strength*

The compressive strength machine was used to test a cube with a particular load until it failed. Table 3 shows the compressive strength results acquired from the experiment.

The compressive strength and percentage of sawdust substitution are proportional in Fig. 4. This experiment showed that the percentage of sawdust substitution has an inverse relationship with concrete compressive strength. Figure 4 shows that the strength of the structure increases with age when sawdust replaces fine aggregate at any proportion. The interaction between cement and water continues, resulting in the formation of more C-S-H gel for bonding (Sekkal & Zaoui, 2017). Furthermore, sawdust-containing concrete mixing ratios of 5%, 10%, 15%, and 20% showed a decrease in compressive strength across the board, regardless of age. The concrete with 5% sawdust addition achieved the maximum strength of 17.2 MPa in this situation. However, compared to the zero replacement of regular concrete, the strength achieved is lower (32.79 MPa). The lowering of concrete strength might be due to the presence of lignin in sawdust, which slows the hydration of cement (Vaickelionis & Vaickelioniene, 2006).

3.3 *Density*

By weighing and calculating the volume, the density of concrete was calculated. Table 4 shows the average density of sand replacement percentages.

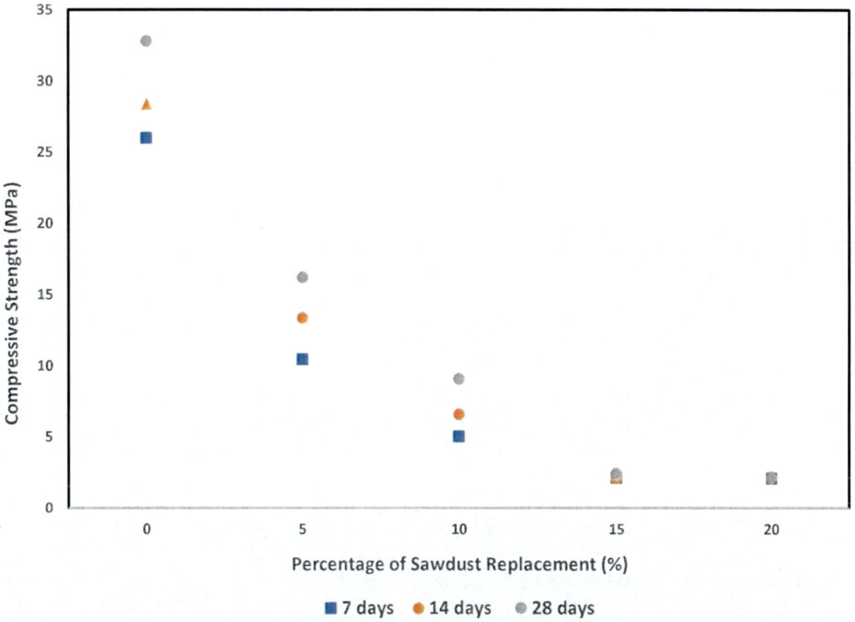

Fig. 4 Compressive strength with different percentages of sawdust as a replacement

Table 4 Density for each percentage of sawdust as a replacement

Percentage replacement (%)	Density (kgm^{-3})		
	7	14	28
0	2410	2420	2445
5	2243	2246	2259
10	2029	2052	2126
15	1921	1952	2004
20	1752	1820	1868

Figure 5 shows a plot of concrete density with sawdust substitution in percentage in various batches. Figure 5 shows that as the amount of sawdust in concrete increases, the density of the concrete decreases. According to the experimental results, the density of concrete fell from 244 to 1868 kgm^{-3}, as the percentage of sawdust was increased from 0 to 20%. Sand has a higher density than sawdust. As a result, when equal weights of sand and sawdust were substituted for sand, the mass of concrete decreased, lowering the density (Dash et al., 2016; Hafidh et al., 2021; Munir et al., 2020; Nanayakkara & Xia, 2019). Furthermore, lightweight concrete is defined as concrete that contains 20% sawdust and has a density of 1868 kgm^{-3} or less than 2000 kgm^{-3}.

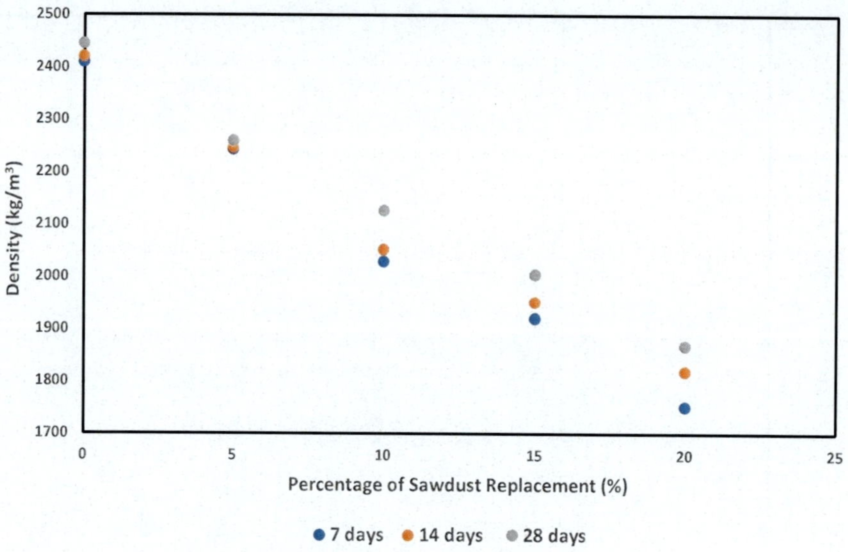

Fig. 5 Density with the percentage of sawdust as a replacement

3.4 Morphological Properties

Figure 6a–d show SEM images of the 5% sawdust replacement in concrete at 100×, 250×, 500×, and 1000× magnifications, respectively. As can be observed, the wood by-product has an uneven fibrous morphology that may be ideal for cement adhesion. A short continuous fibre-like layer is found, extremely carbonaceous in origin and contains a high carbon content (Cheah & Ramli, 2011; Gil et al., 2017). The micrographs show that the sawdust surface is free of substantial damage and that the material is evenly distributed throughout the concrete. However, the flaky structure of cement is noticeable due to the brittle structure of concrete.

3.5 Elemental Properties

Table 5 and Fig. 7 shows the EDS/EDX plot and element contents for 5% sawdust replacement in concrete. The elements present in the sample, according to Table 5, were oxygen (O), calcium (Ca), carbon (C), silicon (Si), iron (Fe), and aluminium (Al). Table 5 shows that oxygen mass percentage has the highest content, followed by

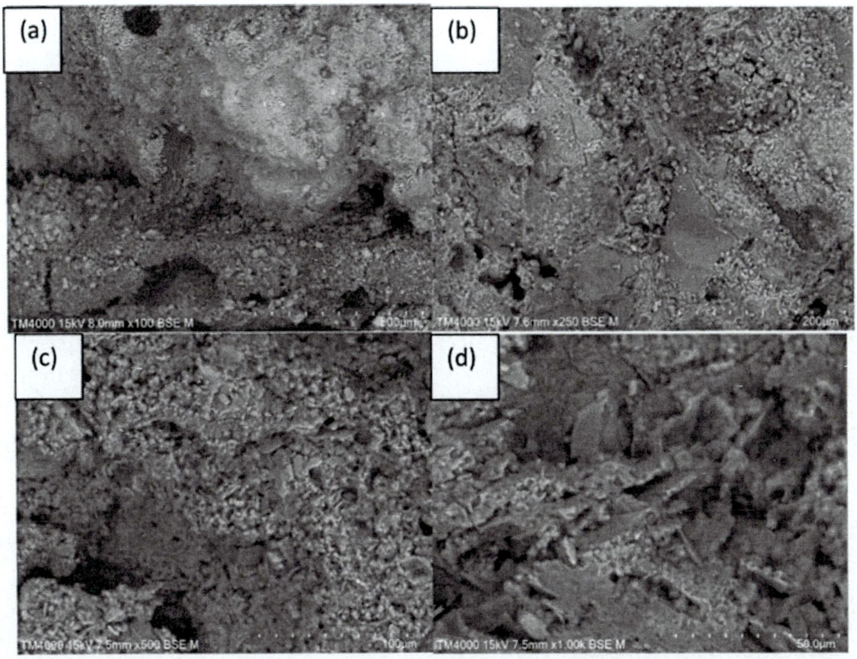

Fig. 6 SEM images for 5% sawdust replacement in concrete at magnification of **a** 100×, **b** 250× , **c** 500**x**, and **d** 1000×

Table 5 Element composition of 5% sawdust replacement in concrete

Element	Mass normal (%)	Atom (%)
O	56.49	67.46
Ca	23.38	11.15
C	9.02	14.35
Si	7.82	5.32
Fe	1.65	0.56
Al	1.64	1.16
Sum	100.00	100.00

calcium and carbon. It indicates that a chemical reaction occurred during the mixing of the paste combination with hydraulic reactivity and sawdust, resulting in calcium carbonate ($CaCO_3$) bonding with the sawdust (Cheah & Ramli, 2011; Gil et al., 2017). $CaCO_3$ improves early strength by having an accelerator effect and a high rate of hydration and causing the concrete to solidify more quickly and be absorbed more easily by the sawdust (Cheah & Ramli, 2011; Gil et al., 2017). Appropriate mixing ratio and chemical substance utilization to establish suitable reactivity, as demonstrated by SEM morphological images, generate a strong link between each material used in the sample combination.

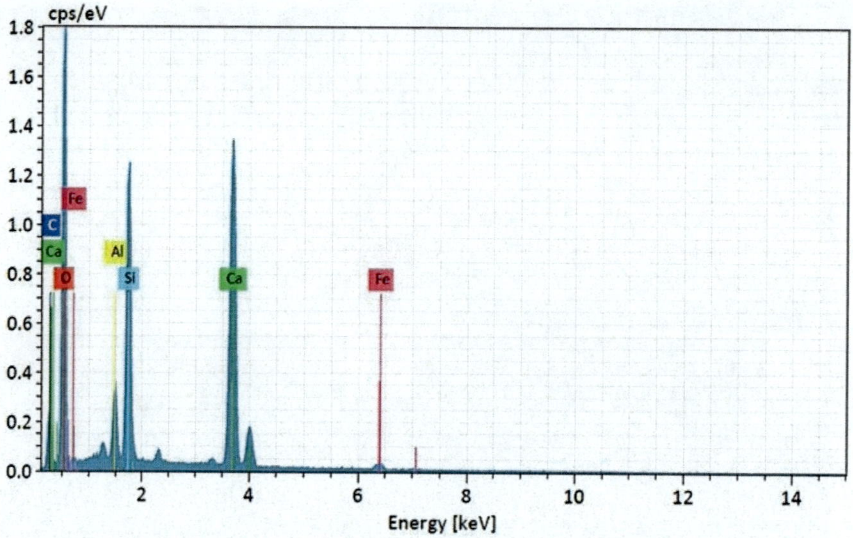

Fig. 7 EDS/EDX graph for 5% sawdust replacement in concrete

3.6 *Infrared Spectral Properties*

Figure 8 shows the FTIR spectrum of 5% sawdust replacement in concrete. The O-H bond generated minor peaks at 3200–3600 cm^{-1} due to the hydraulic of water bond and moisture retained by both cement and wood. Peaks at 1699.29 cm^{-1} and 1514.12 cm^{-1} were attributable to C-H bending and stretching, respectively. At the same time, the 576.72 cm^{-1} peak was due to C-O stretching (Ahmed et al., 2018). It

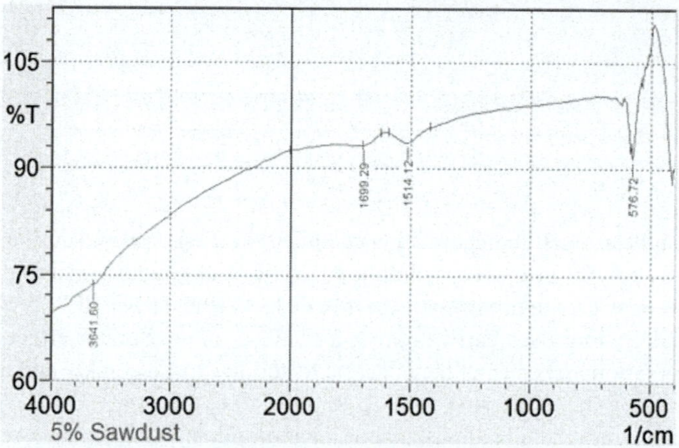

Fig. 8 FTIR spectra of 5% sawdust replacement in concrete

shows the presence of cellulose, hemicellulose, lignin and various hydroxyl groups such as tannins or other compounds that indicated the presence of wood (Sidiras et al., 2011).

4 Application of Sawdust in Concrete

Sawdust's physical and chemical qualities vary greatly based on various circumstances, most notably the kind of wood used. The porous nature of sawdust particles absorbs most of the water, leaving insufficient water for cement hydration. It is also assumed that if sawdust particles are in a saturated surface dry state, the hydration process of the cement will be aided by reducing the requirement for curing because cement particles gather water accumulated in sawdust particles. Sawdust may be used as an alternative to fine aggregate in the manufacture of concrete. Because of the considerable quantity of bark that might impair the setting and hydration of cement, sawdust should be cleansed and cleaned before use as a concrete ingredient. Sawdust concrete combines sawdust, sand, coarse aggregate, and water. Sawdust concrete is low in weight and provides enough thermal insulation. Sawdust has been used in the construction sector to create sawdust concrete, composed of Portland cement, sand, sawdust, and water. This kind of concrete has been shown to bind effectively with regular concrete (Gopinath et al., 2015).

5 Conclusion

The following conclusions are drawn from the results and observations of the experiment conducted in this study on the performance of sawdust as a substitute for sand in concrete. Sawdust can substitute sand in concrete, which might help build construction. As the proportion of sawdust substitution increases, the workability of the concrete mixture decreases. As the sawdust content rises, the water-to-cementitious ratio may be adjusted to produce the best compressive strength. The optimal compressive strength of sawdust concrete is 5% sawdust inclusive, which achieves 17.2 MPa, meeting the requirements for lightweight structural concrete (17 MPa). Sawdust concrete has a lower weight than regular concrete, making it more cost-effective and suitable for use as a lightweight building material. More research needs to be done to determine the flexural and tensile strength of sawdust concrete.

References

Abdel-Shafy, H. I., & Mansour, M. S. M. (2018). Solid waste issue: Sources, composition, disposal, recycling, and valorization. *Egyptian Journal of Petroleum, 27*(4), 1275–1290. https://doi.org/10.1016/j.ejpe.2018.07.003

Abdullahi, A., Abubakar, M., & Afolayan, A. (2013). Partial replacement of sand with sawdust in concrete production. In 3rd Biennial Engineering Conference (pp. 1–6).

Ahmed, A. (2021) Conversion of plastic and sawdust wastes into construction materials. *IOSR: Journal of Environment Science, Toxicology and Food Technology, 15*(2), 57–62. https://www.iosrjournals.org/iosr-jestft/papers/Vol15-Issue2/Series-1/G1502015762.pdf

Ahmed, W., Khushnood, R. A., Memon, S. A., Ahmad, S., Baloch, W. L., & Usman, M. (2018). Effective use of sawdust for the production of eco-friendly and thermal-energy efficient normal weight and lightweight concretes with tailored fracture properties. *Journal of Cleaner Production, 184*(1), 1016–1027. https://doi.org/10.1016/j.jclepro.2018.03.009

Ahuja, D., & Tatsutani, M. (2009). Sustainable energy for developing countries. *Surveys and Perspective Integrating Environment and Society, 2*(1), 1–16. https://journals.openedition.org/sapiens/823

Akadiri, O. P., Chinyio, E. A., & Olomolaiye, P. O. (2012). Design of a sustainable building: A conceptual framework for implementing sustainability in the building sector. *Buildings, 2*(2), 126–152. https://doi.org/10.3390/buildings2020126

Ambiga, K., & Meenakshi, P. (2015). Studies on strength of concrete by partial replacement of sand with sawdust. *International Journal of Advanced Engineering Research and Studies, 20*(20), 1–3. https://www.technicaljournalsonline.com/ijaers/VOL%20IV/IJAERS%20VOL%20IV%20ISSUE%20III%20APRIL%20JUNE%202015/630.pdf

ASTM E1252-98. (2013). Standard practice for general techniques for obtaining infrared spectra for qualitative analysis. ASTM International, West Conshohocken, PA. https://doi.org/10.1520/E1252-98R13E01

ASTM E1508-12. (2019). Standard guide for quantitative analysis by energy-dispersive spectroscopy. ASTM International, West Conshohocken, PA. https://doi.org/10.1520/E1508-12AR19

ASTM C1723-16. (2016). Standard guide for examination of hardened concrete using scanning electron microscopy. ASTM International, West Conshohocken, PA. https://doi.org/10.1520/C1723-16

ASTM E168-16. (2016). Standard practices for general techniques of infrared quantitative analysis. ASTM International, West Conshohocken, PA. https://doi.org/10.1520/E0168-16

ASTM C142/C142M-17. (2017). Standard test method for clay lumps and friable particles in aggregates. ASTM International, West Conshohocken, PA. https://doi.org/10.1520/C0142_C0142M-17

Asuzu, C. C., Okereke, C. D., Asoegwu, S. N., & Nwachukwu, P. I. (2017). Suitability of agricultural waste product (palm kernel shell) as coarse aggregate in concrete: A review. *International Journal of Agriculture, Environment and Bioresearch, 2*(2), 234–240. https://ijaeb.org/uploads/AEB_02_28.pdf

Awal, A. S. M. A., Mariyana, A. A. K., & Hossain, M. Z. (2016). Some aspects of physical and mechanical properties of sawdust concrete. *International Journal of GEOMATE, 10*(21), 1918–1923. https://www.geomatejournal.com/sites/default/files/articles/1918-1923-5304-Awal-May-2016-c1.pdf

Bedi, R., Chandra, R., & Singh, S. P. (2013). Mechanical properties of polymer concrete. *Journal of Composites, 2013*(1), 1–12. https://doi.org/10.1155/2013/948745

Boob, T. N. (2014). Performance of saw-dust in low cost sandcrete blocks. *Americal Journal of Engineering Research, 3*(4), 197–206.

BS EN 12390-3. (2019). Testing hardened concrete. Compressive strength of test specimens. British Standard Institution, London, England.

BS EN 12390-7. (2009). Testing hardened concrete. Density of hardened concrete. British Standard, London.

BS EN 196-1. (2016). Methods of testing cement. Determination of strength. British Standard, London.

Cheah, C. B., & Ramli, M. (2011). The implementation of wood waste ash as a partial cement replacement material in the production of structural grade concrete and mortar: An overview. *Resources, Conversation and Recycling, 55*(7), 669–685. https://doi.org/10.1016/j.resconrec.2011.02.002

Chel, A., & Kaushik, G. (2018). Renewable energy technologies for sustainable development of energy efficient building. *Alexandria Engineering Journal, 57*(2), 655–669. https://doi.org/10.1016/j.aej.2017.02.027

Chowdhury, S., Mishra, M., & Suganya, O. (2015). The incorporation of wood waste ash as a partial cement replacement material for making structural grade concrete: An overview. *Ain Shams Engineering Journal, 6*(2), 429–437. https://doi.org/10.1016/j.asej.2014.11.005

Collivignarelli, M. C., Cillari, G., Ricciardi, P., Miino, M. C., Torett, V., Rada, E. C., & Abba, A. (2020). The production of sustainable concrete with the use of alternative aggregates: A review. *Sustainability, 12*(19), 7903. https://doi.org/10.3390/su12197903

Danso, H., & Obeng-Ahenkora, N. K. (2018). Major determinants of prices increase of building materials on ghanaian construction market. *Open Journal of Civil Engineering, 8*(2), 142–154. https://doi.org/10.4236/ojce.2018.82012

Dash, M. K., Patro, S. K., & Path, A. K. (2016). Sustainable use of industrial-waste as partial replacement of fine aggregate for preparation of concrete—A review. *International Journal of Sustainable Built Environment, 5*(2), 484–516. https://doi.org/10.1016/j.ijsbe.2016.04.006

Gana, A. J., Asebiomo, J., & Joy, O. (2018). Evaluation of kernel shell as construction materials for low cost houses. *International Journal of Environmental Studies and Safety Research, 3*(3), 6–32.

Ganiron, J. T. U. (2014). Effect of sawdust as fine aggregate in concrete mixture for building construction. *International Journal of Advanced Science and Technology, 63*(1), 73–82. https://doi.org/10.14257/ijast.2014.63.07

Gil, H., Ortega, A., & Perez, J. (2017). Mechanical behavior of mortar reinforced with sawdust waste. *Procedia Engineering, 200*(1), 325–332. https://doi.org/10.1016/j.proeng.2017.07.046

Gopinath, K., Anuratha, K., Harisundar, R., & Saravanan, M. (2015) Utilization of saw dust in cement mortar & cement concrete. *International Journal of Scientific & Engineering Research, 6*(10), 665–682. https://www.ijser.org/researchpaper/Utilization-of-Saw-Dust-in-Cement-Mortar-Cement-Concrete.pdf

Hafidh, S. A., Abdullah, T. A., Hashim, F. G., & Mohmoud, B. K. (2021). Effect of adding sawdust to cement on its thermal conductivity and compressive strength. *IOP Conference Series: Materials Science and Engineering, 1094*, 012047. https://doi.org/10.1088/1757-899X/1094/1/012047

Imbabi, M. S., Carrigan, C., & McKenna, S. (2012). Trends and developments in green cement and concrete technology. *International Journal of Sustainable Built Environment, 1*(2), 194–216. https://doi.org/10.1016/j.ijsbe.2013.05.001

IS 10262. (2009). Concrete mix proportioning- guidelines. Indian Standards, New Delhi.

IS 456. (2000). Plain and reinforced concrete – code of practice. Indian Standards, New Delhi.

IS 383. (2003). Specification for coarse and fine aggregates from natural sources for concrete. Indian Standards, New Delhi.

Kaniapan, S., Hassan, S., Ya, H., Nesan, K. P., & Azeem, M. (2021). The utilisation of palm oil and oil palm residues and the related challenges as a sustainable alternative in biofuel, bioenergy, and transportation sector: A review. *Sustainability, 13*(6), 3110. https://doi.org/10.3390/su13063110

Mangi, S. A., Jamaluddin, N. B., Siddiqui, Z., Memon, S. A., & Ibrahim, M. H. B. (2019). Utilization of sawdust in concrete masonry blocks: A review. *Mehran University Research Journal of Engineering & Technology, 38*(2), 487–494. https://doi.org/10.22581/muet1982.1902.23

Mannan, M. A., & Ganapathy, C. (2002). Engineering properties of concrete with oil palm shell as coarse aggregate. *Construction and Building Materials, 16*(1), 29–34. https://doi.org/10.1016/S0950-0618(01)00030-7

Marveas, C. (2020). Production of sustainable construction materials using agro-wastes. *Materials, 13*(2), 262. https://doi.org/10.3390/ma13020262

Munir, A., Memon, B. A., & Oad, M. (2020). Impact of cement replacement by sawdust ash on workability and compressive strength of concrete. *Quest Research Journal, 18*(1), 35–50.

Mwango, A., & Kambole, C. (2019). Engineering characteristics and potential increased utilisation of sawdust composites in construction—A review. *Journal of Building Construction and Planning Research, 7*(3), 59–88. https://doi.org/10.4236/jbcpr.2019.73005

Nayakkara, O., & Xia, J. (2019) Mechanical and physical properties of mortar of partially replaced fine aggregates with sawdust. *3rd International Conference on Bio-Based Building Materials, 37*(2), 308–313.

Osei, D. Y., & Jackson, E. N. (2016). Compressive strength of concrete using sawdust as aggregate. *International Journal of Scientific & Engineering Research, 7*(4), 1349–1353. https://www.ijser.org/researchpaper/Compressive-Strength-of-Concrete-Using-Sawdust-as-Aggregate.pdf

Oyedepo, O. J., Oluwajana, S. D., & Akande, S. P. (2014). Investigation of properties of concrete using sawdust as partial replacement for sand. *Civil and Environmental Research, 6*(2), 35–42.

Priyadharshini, B., Nirmala, R., Eshanthini, & Kumaar, M. C. (2021). Experimental study on concrete with partial replacement of coarse aggregate with cashew shells. *Journal of Physics: Conference Series, 1770*(1), 012036. https://doi.org/10.1088/1742-6596/1770/1/012036

Prusty, J. K., Patro, S. K., & Basarkar, S. S. (2016). Concrete using agro-waste as fine aggregate for sustainable built environment—A review. *International Journal of Sustainable Built Environment, 5*(2), 312–333. https://doi.org/10.1016/j.ijsbe.2016.06.003

Reddy, B. D., Jyothy S. A., & Shaik, F. (2014). Experimental analysis of the use of coconut shell as coarse aggregate. *IOSR Journal of Mechanical and Civil Engineering, 10*(6), 6–13. http://www.iosrjournals.org/iosr-jmce/papers/vol10-issue6/B01060613.pdf

Sekkal, W., & Zaoui, A. (2017). Enhancing the interfacial bond strength of cement nanocomposite with carbonate nanostructure. *Composites Part b: Engineering, 124*(1), 111–119. https://doi.org/10.1016/j.compositesb.2017.05.057

Sidiras, D., Batzias, F., Shroeder, E., Ranjan, R., & Tsapatsis, M. (2011). Dye adsorption on autohydrolyzed pine sawdust in batch and fixed-bed systems. *Chemical Engineering Journal, 171*(3), 883–896. https://doi.org/10.1016/j.cej.2011.04.029

Sojobi, A. O., & Alavi, A. H. (2016). Evaluation of the performance of eco-friendly lightweight interlocking concrete paving units incorporating sawdust wastes and laterite. *Cogent Engineering, 3*(1), 1255168. https://doi.org/10.1080/23311916.2016.1255168

Soliman, N. K., & Moustafa, A. F. (2020). Industrial solid waste for heavy metals adsorption features and challenges; A review. *Journal of Materials Research and Technology, 9*(5), 10235–10253. https://doi.org/10.1016/j.jmrt.2020.07.045

Suliman, N. H., Razak, A. A. B., Mansor, H., Alisibramulisi, A., & Amin, N. M. (2019). Concrete using sawdust as partial replacement of sand: Is it strong and does not endanger health? *MATEC Web of Conference, 258*(1), 01015. https://doi.org/10.1051/matecconf/201925801015

Turgut, P., & Algin, H. M. (2007). Limestone dust and wood sawdust as brick material. *Building and Environment, 42*(9), 3399–3403. https://doi.org/10.1016/j.buildenv.2006.08.012

Udoeyo, F. F., & Dashibil, P. U. (2002). Sawdust ash as concrete material. *Journal of Materials in Civil Engineering, 14*(2), 1–10. https://doi.org/10.1061/(ASCE)0899-1561(2002)14:2(173)

Ugwu, J. N. (2019) Saw dust as full replacement of fine aggrgate in lightweight concrete: Any comparable strength? *The International Journal of Engineering and Sciences, 8*(10) 9–11. http://www.theijes.com/papers/vol8-issue10/Series-1/B0810010911.pdf

Vaickelionis, G., & Vaickelioniene, R. (2006). Cement hydration in the presence of wood extractives and pozzolan mineral additives. *Ceramics-Silikáty, 50*(2), 115–122. https://www.ceramics-silikaty.cz/2006/pdf/2006_02_115

Yalley, P. O., & Sam, A. (2018). Effect of sand fines and water/cement ratio on concrete properties. *Civil Engineering Research Journal, 4*(3), 1–7. https://doi.org/10.19080/CERJ.2018.04.555636

Plastic Waste as Fine Aggregate for Sand Filler Replacement in Concrete

Chin Mei Yun, Md Rezaur Rahman, Durul Huda, Kuok King Kuok, Amelia Chai Pei Sze, Dahlia Chan Xin Lin, and Muhammad Khusairy Bin Bakri

Abstract Since its invention in 1907, plastic has been widely utilized worldwide. As plastics were thrown away after a single use, a severe problem arose, resulting in substantial pollution and contamination of the environment. Plastic bottles accounted for a significant portion of the plastic trash disposal. Plastic bottles are frequently made from waste polyethylene terephthalate (WPET). As a result, PET plastic bottles were employed as a by-product in this experiment since they are light in weight and, more significantly, help decrease trash pollution. The purpose of this study is to investigate the properties of concrete by partially replacing sand with plastic waste as fine aggregates, with the goal of this experiment being to determine the optimum percentage of sand replacement with plastic while maintaining compressive strength or achieving similar strength within the grade used when compared to reference concrete. 7 batches of 6 cubes each have been cast. One batch served as a control, three batches were cast without the addition of superplasticizer, and three more batches were cast with the optimal percentage of PET plastic replacement after the addition of superplasticizer. The subject of this investigation was the compressive strength test using 0%, 5%, 10%, and 15% PET plastic substituted with sand in concrete. After 7 days and 28 days of curing, a compressive strength test was performed. The plastic concrete's maximum compressive strength was 27.6 MPa, which appears to be somewhat lower than the control sample. When a slight variation in strength is detected, the results are acceptable. The slightly lower strength obtained might be because of plastic's hydrophobic material. Therefore, a superplasticizer was added to improve the workability and compressive strength of the optimal plastic concrete. Naphthalene sulfonate formaldehyde was employed as a superplasticizer. Further research

C. M. Yun · K. K. Kuok · A. C. P. Sze · D. C. X. Lin
Faculty of Engineering, Computing and Science, Swinburne University of Technology Sarawak Campus, Jalan Simpang Tiga, 93400 Kuching, Sarawak, Malaysia

M. R. Rahman (✉) · M. K. B. Bakri
Faculty of Engineering, Universiti Malaysia Sarawak, Jalan Datuk Mohammad Musa, 94300 Kota Samarahan, Sarawak, Malaysia
e-mail: rmrezaur@unimas.my

D. Huda
Department of Mechanical Engineering and Product Design Engineering, Swinburne University of Technology, Hawthorn, VIC 3122, Australia

M. R. Rahman et al. (eds.), *Waste Materials in Advanced Sustainable Concrete*, Engineering Materials, https://doi.org/10.1007/978-3-030-98812-8_9

was carried out to understand better the characteristics of partial sand replaced with plastic, with or without a superplasticizer.

Keywords PET plastic · Reference concrete · Superplasticizer · Compressive strength

1 Introduction

Human development has resulted in demand for extensive infrastructure development (Kniivila, 2007). The continual extraction of sand from riverbeds depletes resources in the not-too-distant future (Ashraf et al., 2011; Bravard & Gaillot, 2013; Gavriletea, 2017; Gondo et al., 2019). The increased demand for sand increased material costs (Elango & Ashok Kumar, 2018). Another problem that disturbed the public, which gained attention, was the uncontrolled waste growth towards the garbage disposal to the landfills (Agarwal et al., 2015; Ferronato & Torretta, 2019; McAllister, 2015; Thompson et al., 2009). Plastic garbage is one of the most significant environmental waste contributors (Alabi et al., 2019; Chen et al., 2021). Although plastic makes life easier for humans, it is renowned for its low biodegradability and takes thousands of years to disintegrate (Thompson et al., 2009). According to Balasegaram (2018), Malaysia's contribution to global plastic trash is in the top eight. Due to a shortage of recycling facilities, Malaysia produced half a million tons of plastic trash in 2010, and this number continues to rise each year (Jereme et al., 2015). By combining these two concerns, it is critical to investigate the options for replacing sand with plastic to decrease pollution and avoid sand depletion in the future while also ensuring that the replacement concrete maintains its strength (Batayneh et al., 2007).

Few past studies have investigated the possibility of partially replacing sand with discarded PET plastics (Almeshal et al., 2020; Mustafa et al., 2019). The water to cement ratio, the types of plastics used, the sizes of plastics, and other variables have all been documented as contributing to the decline in concrete compressive strength (Umasabor & Daniel, 2020). The possibility of replacing tiny angular particles onto concrete with three different types of water to binder ratios was investigated by Al-Manaseer and Dalal (1997). It was discovered that when the plastic content and water to binder ratio rise, the compressive strength falls (Yurdakul et al., 2014). Poor bonding between the plastic and the cement paste was blamed for losing strength. Choi et al. (2005) discovered that when extra water did not react with cement, a more excellent water content ratio resulted in a lower compressive strength value. As a result of water evaporation, capillaries have been reduced in diameters. Figure 1 shows the graph of loss in compressive strength with an increase in the proportion of plastic.

Frigione (2010) investigated the impact of partial sand substitution on the mechanical characteristics of concrete. PET polymers that were granulated and graded compared to river sand were utilized. The results showed that the increase in plastic substitution caused a decrease in compressive strength (Hama, 2021; Lee et al., 2019;

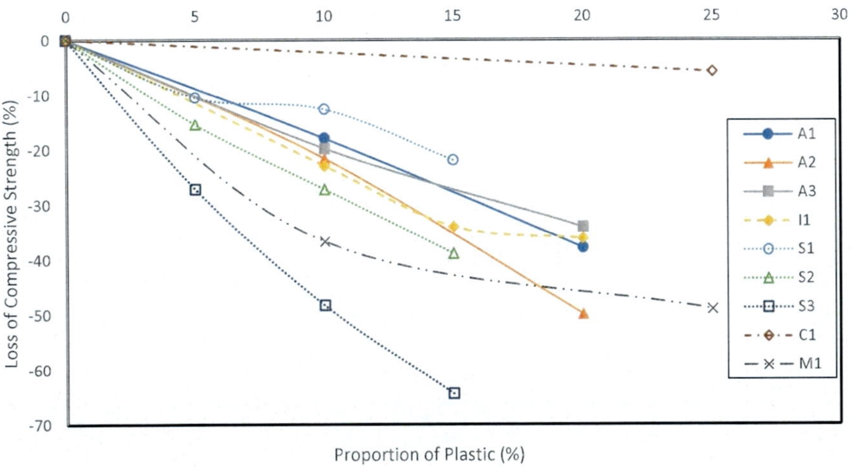

	A = Albano et al. (2009)	I = Ismail and Al-Hashmi (2008)	S = Saikia and de Brito (2014)	C = Choi et al. (2005)	M = Al-Manaseer and Dalal (1997)
1	All small particles	PET and polystyrene	PP - cylindrical pellets	PET coated in GBFS	Angular particles
2	All large particles	N/A	PF - flaky fine particles	N/A	N/A
3	50/50 mix of particles	N/A	PC - large coarse particles	N/A	N/A

Fig. 1 Loss in compressive strength of concrete and increase in the proportion of plastic (Thorneycroft et al., 2018)

Nursyamsi & Adil, 2021). The best proportion of plastic replacement was found to be 5%. The achieved compressive strength has been just 2% lower than the reference concrete. Although the achieved strength was lower than the reference standard for concrete, it was still within the concrete grade M20 (Dash et al., 2017; Prusty et al., 2016). As there is an increase in plastic material replacement, larger plastic size and higher water content ratio resulted in shallow strength values, according to Frigione (2010) and Albano et al. (2009). The reason offered may be due to honeycomb development in the concrete (Stefan et al., 2018). Figures 2 and 3 shows the compressive strength of a concrete-PET blend with different curing age at 0.5 and 0.6 water/cement ratio, respectively.

Ismail and Al-Hashmi (2008) found that replacing small shredded WPET with sand reduced the cost of concrete and solved solid waste problems caused by the production of plastics, but that the compressive strength and slump test were reduced as the amount of plastic placed in each curing age was increased (Ismail & Al-Hashmi, 2008). It was revealed that this was due to the hydrophobic nature of plastics, which prevented cement hydration (Zulkernain et al., 2021). In addition, the WPET sizes had a significant influence in achieving the results. Even though the value of droop has decreased, plastic is still appropriate for usage because of its ease of workability. Figures 4 and 5 show the compressive strength and slump. Therefore, Ferreira et al.

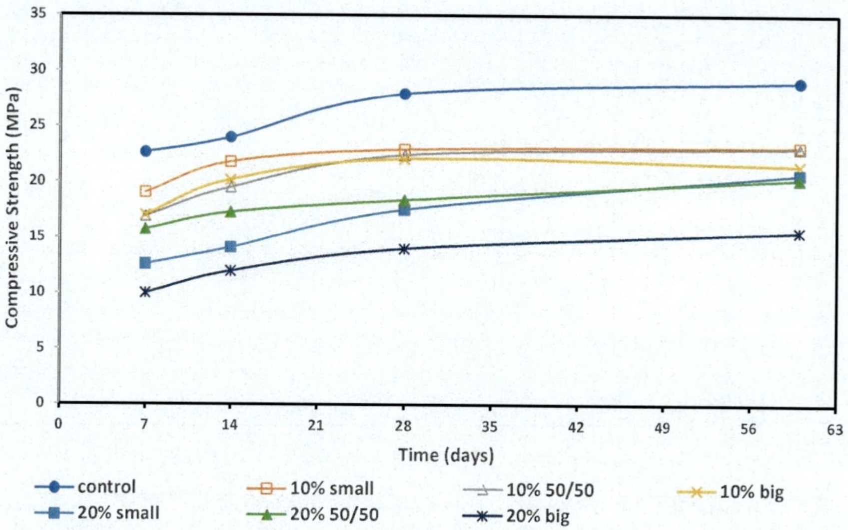

Fig. 2 Compressive strength of a concrete-PET blend with different curing ages at 0.5 water/cement ratio (Albano et al., 2009)

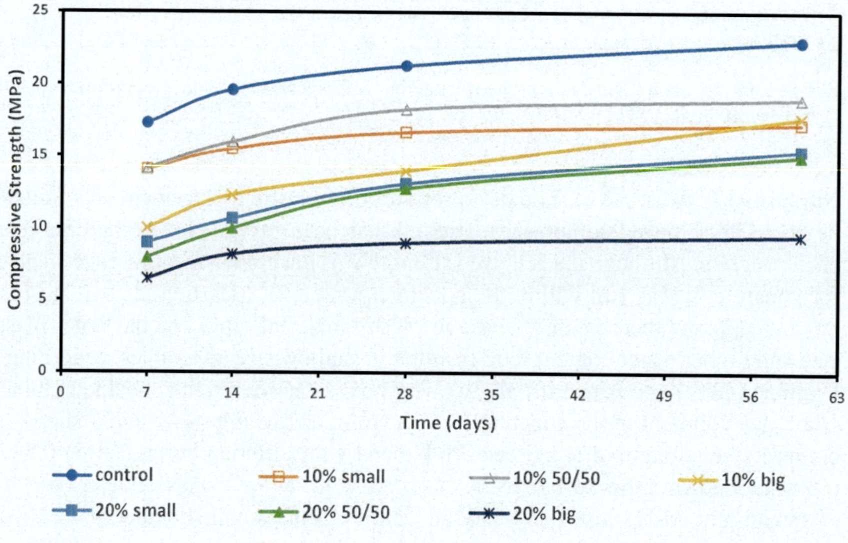

Fig. 3 Compressive strength of a concrete-PET blend with different curing ages at 0.6 water/cement ratio (Albano et al., 2009)

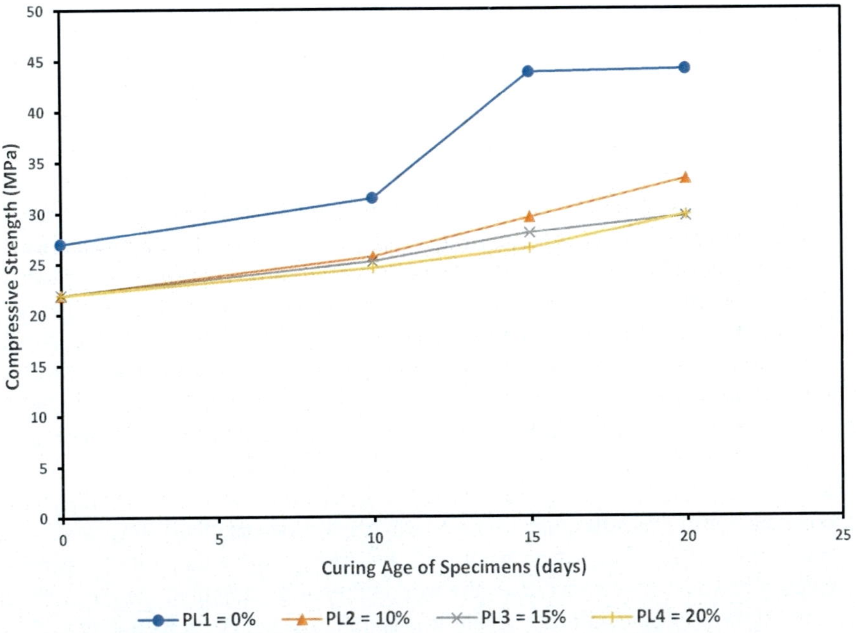

Fig. 4 Compressive strength of waste plastic concrete (Ismail & Al-Hashmi, 2008)

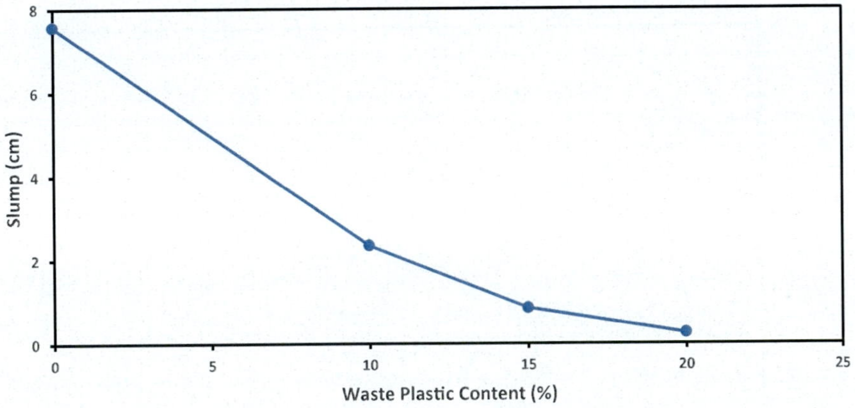

Fig. 5 The slump of waste plastic concrete

[8] demonstrated that the curing age of concrete with a plastic replacement did not affect performance; instead, the proportion of plastic replacement affected the strength values in concrete.

According to Rai et al. (2012), as the volume and size of plastic trash increased, the compressive strength of concrete dropped during each curing day. The adhesive strength between the cement and the plastic surface tends to deteriorate (Almeshal et al., 2020; Rai et al., 2012). The addition of a superplasticizer to the plastic replacement concrete, on the other hand, improved the strength by 5% (Salem et al., 2016). As fine aggregates, plastic pallets were utilized, with a constant water content ratio of 0.44. It was discovered that adding superplasticizer to the mixture enhanced compressive strength and slump value (Musbah et al., 2019). The assertion is also corroborated by Patanaik and Rath (2018), who found that adding superplasticizer to concrete enhanced the strength by up to 1.5% and decreased strength after another 1.5% was added. The water content ratio has been reported to be maintained. The decrease in compressive strength of concrete following the addition of 3% superplasticizer was explained by the mix becoming harsh. Figure 6 shows the slump of the concrete, while Figs. 7 and 8 show the compressive strength of concrete with and without plasticizer.

In addition, there was a little study on the use of superplasticizers in plastic concrete. It was discovered that a proportion of plastic replacement with sand of 5% was optimal. The concrete created had a little lower compressive strength than normal concrete, but it was still quite strong (Awoyera & Adesina, 2020; Bag et al., 2020; Balea et al., 2021). The research revealed that adding a little amount of superplasticizer to plastic concrete increased the workability and strength of the concrete until 1.5% of the total cementitious weight was applied (Akije, 2019; Koting et al., 2007; Salem et al., 2016). At more than 1.5% superplasticizer was applied, the strength of the material decreased (Bethe et al., 2019; Carabba et al., 2016). Even though

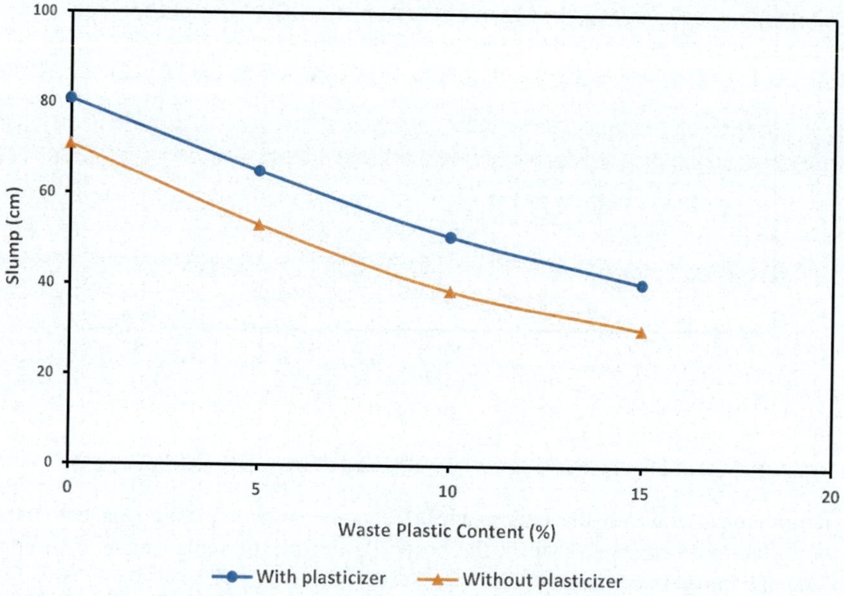

Fig. 6 The slump of concrete mixes with varying percentages of waste plastic (Rai et al., 2012)

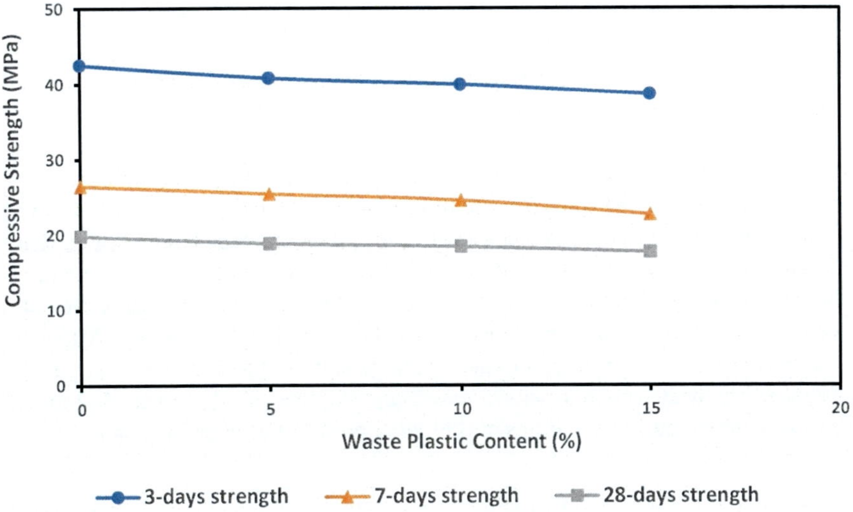

Fig. 7 Compressive strength of concrete with varying percentage of plastic waste without superplasticizer (Rai et al., 2012)

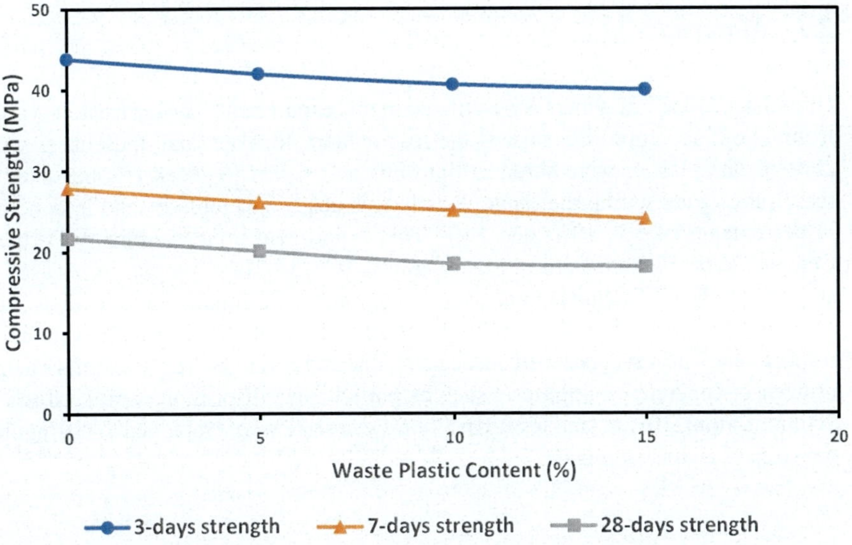

Fig. 8 Compressive strength of concrete with varying percentage of plastic waste with superplasticizer (Rai et al., 2012)

previous studies have been conducted, there is still a knowledge gap about the types and sizes of plastics used.

The first and second phases of the experimental procedure were separated. The first stage partially replaced the sand with various percentages of plastics, followed by adding a superplasticizer to the optimally substituted plastic concrete. To investigate the variations in concrete strength, before and after adding superplasticizer, two phases were dispersed with four and three batches of the concrete cast, respectively. After replacing sand with waste plastics, no previous studies have found that the strength of the concrete is higher than that of normal concrete. As a result, more research was done in this experiment to see if adding superplasticizer to the optimal replacement plastic concrete might enhance the strength of the concrete. After adding superplasticizer to concrete, the findings were expected to improve its strength and workability. M20 concrete with a cure time of 7 and 28 days was chosen as the reference concrete. The findings of this study might be used in the concrete industry as extra data and references.

2 Methodology

2.1 Materials

Granulated WPET polymers were utilized in this experiment. The dimensions varied from 0.0625 to 2 mm. Because of the resemblance to river sand, these sizes were chosen. The plastics were shredded into bits and sieved to check that the sizes of the plastics were within the range. Plastic was utilized to replace sand in concrete in percentages of 5%, 10%, and 15%. This experiment utilized ordinary Portland Cement Grade 53 from Cahaya Mata Sarawak (CMS). The superplasticizer used in stage two is Kao (Thailand) Industrial's Naphthalene Sulfonate Formaldehyde with Mighty Kao 150. 0.8%, 1.0%, and 1.2% of the total cementitious weight were used, respectively. The test cubes were made using fine and coarse aggregates. Before being utilized in concrete casting, tests such as particle size dispersion were performed. 20 mm, 14 mm, 10 mm, and less than 5 mm sieve sizes were employed to distinguish the sizes of coarse aggregate.

2.2 Sample Mix Proportion

The mix proportion design of concrete was determined using India Standard IS 10262 (2009). As a reference, a concrete sample of grade 20 was employed. The slump value must be between 75 and 100 mm. After 28 days of curing, concrete was projected to reach a strength of 20 to 29 MPa. Table 1 shows the mix proportions and a reference concrete (RC) sample and the mix proportions of testing concrete in kg/m^3.

Table 1 Samples mix proportion

Stage	Samples	Percentage of plastic (%)	Waste plastic (kg/m^3)	Mix proportion	Cement (kg/m^3)	Sand (kg/m^3)	Coarse aggregates (kg/m^3)	Water (kg/m^3)	Superplasticizer (%) of total cementitious weight
1	RC	0	0	1:1.5:3	430	645	1290	214.3	0
	WPET5	5	32.9			613		214.3	0
	WPET10	10	65.7			580		214.3	0
	WPET15	15	97.1			549		214.3	0
2	WPET5 SP0.8	5	32.9			613		171.4	0.8
	WPET5 SP1.0	5	32.9			613		171.4	1.0
	WPET5 SP1.2	5	32.9			613		171.4	1.2

RC means reference concrete, whereas PET5 means polyethylene terephthalate plastic with 5% plastic replacement
SP means superplasticizer, whereas 0.8 is the volume of superplasticizer added to the concrete based on total cementitious weight

2.3 Concrete Preparations, Casting, and Testing

The concrete casting process is separated into two stages: stage 1 and stage 2. This experiment aims to determine the workability and compressive strength of sample concrete and plastic concrete with and without superplasticizers. Before tests were carried out, preparations of materials, fixing of moulds, and screening of aggregates was done to guarantee the concrete casting was carried out smoothly. Seven batches of concrete samples, each with six cubes, were cast and separated into two stages. The first stage primarily consisted of casting RC, PET5, PET10, and PET15 concrete.

In contrast, the second stage consisted of casting PET SP0.8, PET SP1.0, and PET SP1.2 after achieving the optimal proportion of plastic replaced with sand on stage 1. The moulds used were 100 mm × 100 mm × 100 mm in size. After preparing the ingredients, they were placed into the drum mixer, where they were tested for droop and subsequently cast into test cubes if the slump was between 75 and 100 mm. Placed the concrete aside for 24 h ± 8 h after casting. Demold the next day. Samples were then placed in the curing tank for 7 days and 28 days. The samples were tested after curing. Data collected for the compressive strength test were analyzed and compared.

2.4 Workability Test

A slump test was carried out on fresh concrete to ensure the workability of concrete casting. According to ASTM C143/C143M-20 (2020), the slump test was performed.

2.5 Compressive Test

A compressive strength test was carried out using Unit Test Autocon 2000 (Promat (HK) Ltd., Hong Kong) under BS EN 12390-3 (2019). During the compressive strength test, the load was gradually increased at 2.4 ± 0.2 kN/s until specimens failed. The results were averaged and recorded for each concrete specimen from the same batch.

2.6 Scanning Electron Microscope (SEM) and Energy Dispersive X-Ray/Spectroscopy (EDS/EDX)

SEM and EDX/EDS were carried out in compliance with ASTM C1723-16 (2016) and ASTM E1508-12 (2019) standards, respectively. 100×, 250×, 500×, and 10,000× magnifications were used to analyze the samples. Automated software scans and analyses the elemental composition percentages of the samples. The EDX/EDS was conducted on each sample many times at various phases, with the most representative results chosen. Hitachi TM4000Plus Tabletop Microscope (Hitachi Ltd., Tokyo, Japan) with a Quantax75TM Series Energy Dispersive X-Ray Spectrometer was used in the SEM and EDX/EDS investigations of the materials.

2.7 *Fourier Transform Infrared Spectroscopy*

Fourier-transform infrared spectroscopy (IRAffinity-1, Shimadzu Corporation, Kyoto, Japan) was used for the FTIR analysis of the samples. Fourier-transform infrared spectroscopy was conducted according to the ASTM E168-16 (2016) and ASTM E1252-98 (2021) standards for qualitative and quantitative analysis. The spectrum scanning was conducted in the wavenumber range of 4000–400 cm^{-1} for each sample. Fourier-transform infrared spectroscopy utilized the samples' infrared spectrum transmittance and absorption to develop a unique molecular fingerprint spectrum. The test was repeated numerous times for each sample, and the most representative results were selected.

3 Results and Discussion

3.1 *Workability of Concrete*

Slump tests were conducted to investigate the workability of concrete. Slump values were predicted to be between 75 and 100 mm. With an increase in the amount of plastic replacement, the slump value was found to drop. After adding superplasticizer, the slump value appeared to increase with the increase in superplasticizer volume, consistent with previous results that superplasticizer improves concrete workability (Salem et al., 2016). Because superplasticizer acts as a water-reducing agent, it can reduce up to 30% of the water content in concrete, the strength of the concrete increased after it was added (Xun et al., 2020). Another factor for the rise in slump value is the simple workability of WPET plastics, which respond well with superplasticizers (Sahmaran et al., 2013). The graph of samples versus slump value acquired during the experiment is shown in Fig. 9.

3.2 *Compressive Strength*

Figures 10 and 11 demonstrate the compressed strength findings acquired in stages 1 and 2 in this experiment, which were plotted into a graph according to 7 days and

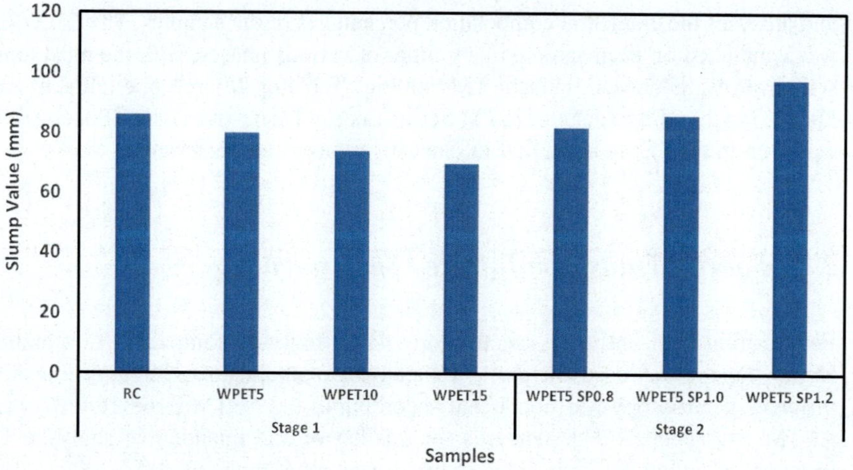

Fig. 9 Samples slump values (mm)

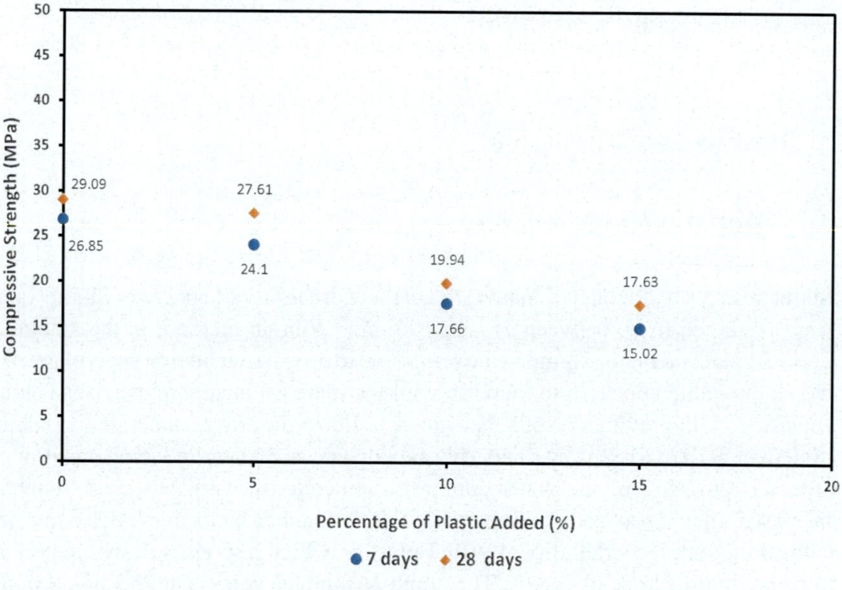

Fig. 10 Compressive strength with different percentages of plasticizer added

28 days curing age. The findings were calculated using the average of at least three samples. Figure 10 shows that the optimal plastic replacement with sand yields a compressive strength of 27.61 MPa and compressive strength of 5%.

Although, compared to ordinary concrete, there is a reduction in concrete strength. On the other hand, the concrete's strength is still graded 20. The results of this

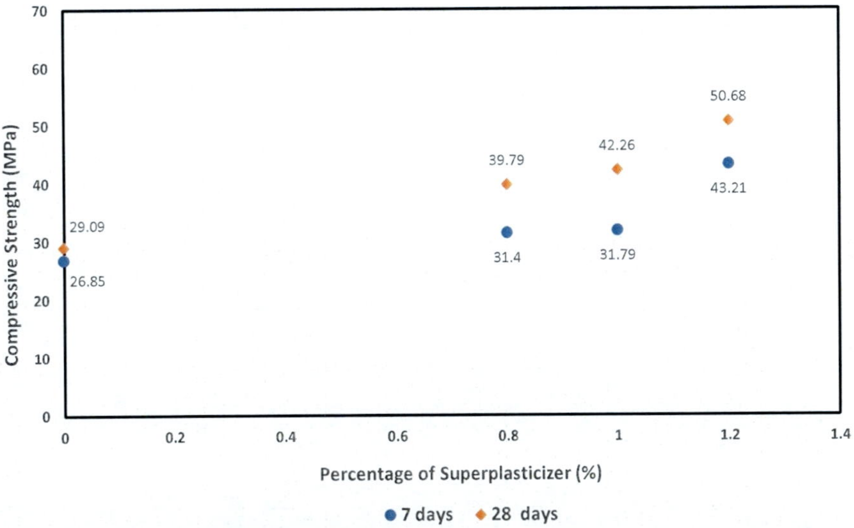

Fig. 11 Compressive strength with different percentages of superplasticizer added to 5% Plastic concrete based on total cementitious weight (%)

experiment are comparable to those of Frigione (2010), who discovered that the optimal plastic substitution is 5% and that there is a slight reduction in strength compared to sample concrete. The decrease in compressive strength after adding 10% and 15% plastic may be explained because plastics are hydrophobic and do not mix with cement paste, as seen in the graph. As a result, when compared to reference concrete, the interfacial transition zone is weaker. Figure 11 shows that when the amount of superplasticizer rises, the compressive strength also increases. With the addition of 1.2% of total cementitious weight, the greatest compressive strength was obtained was 50.68 MPa. As a water lowering agent, the superplasticizer enhanced the strength and workability of the concrete (Olowofoyeku et al., 2019).

3.3 Morphological Properties

Figure 12a–d show morphological images for WPET5 (7 days) sample. The morphological images show that the fine aggregate of plastic waste in the concrete specimen is in the transition zone with cement paste, which shows a tendency to the formation of C–S–H, which is responsible for the hardening of the cement matrix. Brittle crystal structure, a look-alike, was founded due to hydraulic hydration of water correlation with fine aggregate.

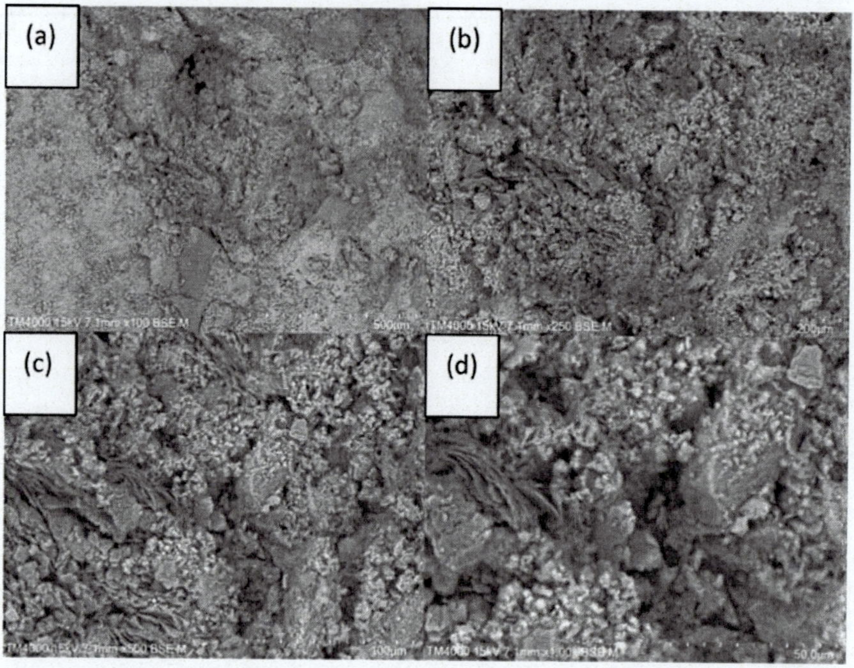

Fig. 12 SEM images for WPET5 (7 days) at magnification of **a** 100×, **b** 250×, **c** 500×, and **d** 1000×

3.4 Elemental Properties

The EDS/EDX plot and element contents for WPET5 (7 days) are shown in Table 2 and Fig. 13. Oxygen (O), calcium (Ca), silicon (Si), carbon (C), aluminium (Al), iron (Fe), and sulfur (S) were found in the sample. Table 2 shows that oxygen, calcium, and carbon have the most significant mass proportions. A chemical reaction happens as the paste, and hydraulic reactivity are combined, resulting in calcium carbonate,

Table 2 Element composition for WPET5 (7 days)

Element	Mass normal (%)	Atom (%)
O	44.75	61.50
Ca	42.00	23.04
Si	5.45	4.27
C	5.18	9.49
Al	1.21	0.99
Fe	0.86	0.34
S	0.56	0.38
Sum	100.00	100.00

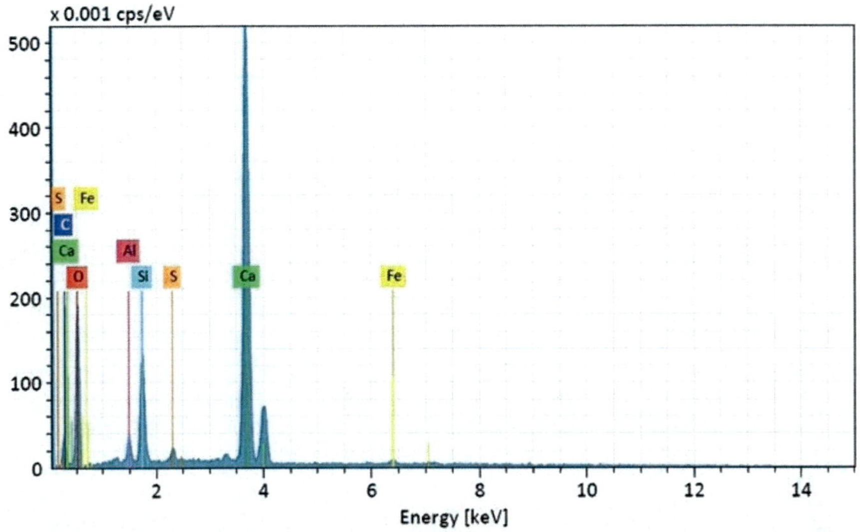

Fig. 13 EDS/EDX graph for WPET5 (7 days)

$CaCO_3$ and C-S.H. $CaCO_3$ enhances early strength by acting as an accelerator and having a high hydration rate, enabling the concrete to harden faster (Allison et al., 2015). Other chemical reactions that revealed the existence of fine plastic aggregate in the concrete were SiO_2, Al_2O_3, C-S.H, and Fe_2O_3 (Allison et al., 2015).

3.5 *Infrared Spectral Properties*

Figure 14 shows the FTIR spectra of WPET5 (7 days). The O-H bond generated a tiny broadband peak at 3400–3600 cm^{-1} due to the hydraulics of water bond and moisture. The peak at 2567.25, 2000.78, and 2025.26 belonged to the carboxyl group (Turku et al., 2016). The peak bands for 1411.89 cm^{-1}, 839.03.72 cm^{-1}, and 588.29 cm^{-1} were due to Si-O-Si and Si-OH stretching. During cement setting, it was observed that Si-OH polymerizes to Si-O-Si (Thoo et al., 2013).

3.6 *Application and Comparison*

Plastic recycling and aggregate are used by replacing fine aggregates in concrete and mortar with plastic powder crushed to less than 2 mm in size. The instances of recycling plastics and utilizing them as fine aggregates are summarised in Table 3. Although concrete contains plastics such as PET, HDPE, LDPE, LLDPE, and PP, most of the plastics utilised are recycled PET bottles. Plastics utilised in

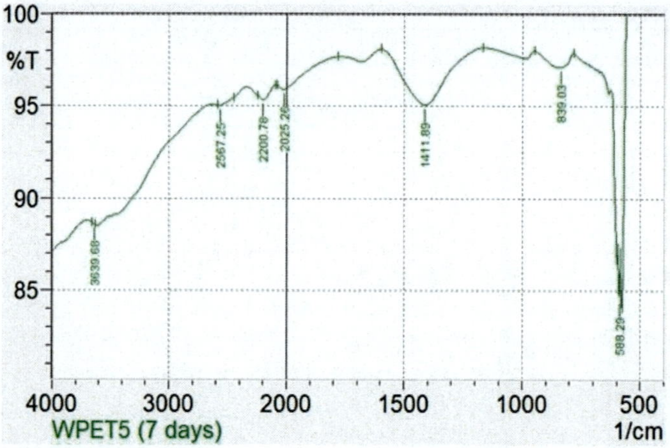

Fig. 14 FTIR for WPET5 (7 days)

Table 3 Types of fine plastic aggregate and applications

Type of plastic	Origin of plastic waste	Recycling procedure	References
High-density polyethylene (HDPE)	Mixed	Shredding	Naik et al. (1996)
Polyethylene terephthalate (PET)	Bottle	Grinding	Choi et al. (2005)
Mix plastic of 80% PET and 20% PS	-	Crushing	Ismail and Al-Hashmi (2008)
PET	Bottle	Grinding	Albano et al. (2009)
PET	Bottle	Crushing	Kou et al. (2009)

lightweight concrete included expanded polystyrene (EPS), PS, polyurethane (PUR), and ethylene-vinyl acetate (EVA), with experiments focusing on recycled EPS foam as a lightweight aggregate. Pulverizing with a shredder is the most common method for converting plastic into aggregate. Pellets made to a specific size by melting were utilised in several experiments (Ferreira et al., 2012; Saikia & de Brito, 2013; Silva et al., 2013).

Furthermore, there have been studies that used grinding and crushing; however, these procedures are not significantly different from grinding and were thus used to regulate the size of the recycled product. Choi et al. (2005) created waste PET lightweight aggregate (WPLA) by combining PET bottles with river sand powder. Algahtani et al. (2017) used compression and heating to create a solid sheet from LLDPE and a red dune sand filler in a 30:70 ratio and then crushed the solid sheet to create aggregate. As a result, washing and grinding were used to prepare plastic as an aggregate in cement-based products.

4 Conclusion

After completing this experiment, it is possible to partially replace the sand with plastics, both with and without using a superplasticizer. The optimal plastic concrete achieved is 5% plastic, with a superplasticizer dose of 1.2% of the total cementitious weight. The increase in concrete strength demonstrated the experiment's feasibility. As a result, it is expected that plastic substitution in concrete employed in the concrete industry will soon lower material costs and minimize pollution.

Acknowledgements The authors would like to thank Swinburne University of Technology Sarawak Campus and Universiti Malaysia Sarawak (UNIMAS) for the collaboration.

References

Agarwal, R., Chaudhary, M., & Singh, J. (2015). Waste management initiatives in India for human well being. *European Scientific Journal, 1*(1), 105–127. https://home.iitk.ac.in/~anubha/H16.pdf

Akije, I. (2019). Characteristic and effects of a superplasticizer quantity variation in some concrete strengths. *Optimization, 38*(1), 81–92. https://doi.org/10.4314/njt.v38i1.11

Alabi, O. A., Ologbonjaye, K. I., Awosolu, O., & Alalade, O. E. (2019) Public and environmental health effects of plastic wastes disposal: A review. *Journal of Toxicology and Risk Assessment, 5*(1), 1–13. https://doi.org/10.23937/2572-4061.1510021

Albano, C., Camacho, N., Hernandez, M., Matheus, A., & Gutierrez, A. (2009). Influence of content and particle size of waste pet bottles on concrete behavior at different w/c ratios. *Waste Management, 29*(10), 2707–2716. https://doi.org/10.1016/j.wasman.2009.05.007

Algahtani, F. K., Ghataora, G., Khan, M. I., & Dirar, S. (2017). Novel lightweight concrete containing manufactured plastic aggregate. *Construction and Building Materials, 148*(1), 386–397. https://doi.org/10.1016/j.conbuildmat.2017.05.011

Allison, P. G., Weiss, C. A., Jr., Moser, R. D., Diaz, A. J., Rivera, O. G., & Holton, S. S. (2015). Nanoindentation and SEM/EDX characterization of the geopolymer-to-steel interfacial transition zone for a reactive porcelain enamel coating. *Composites Part B: Engineering, 78*(1), 131–137. https://doi.org/10.1016/j.compositesb.2015.03.011

Al-Manaseer, A. A., & Dalal, T. R. (1997). Concrete containing plastic aggregates. *Concrete International, 19*(8), 47–52.

Almeshal, A., Tayeh, B. A., Alyousef, R., Alabduljabbar, H., & Mohamed, A. M. (2020). Eco-friendly concrete containing recycled plastic as partial replacement for sand. *Journal of Materials Research and Technology, 9*(3), 4631–4643. https://doi.org/10.1016/j.jmrt.2020.02.090

Ashraf, M. A., Maah, M. J., Yusoff, I., Wajid, A., & Mahmood, K. (2011) Sand mining effects, causes and concerns: A case study from Bestari Jaya, Selangor, Peninsular Malaysia. *Scientific Research and Essays, 6*(6), 1216–1231. https://academicjournals.org/article/article1380635625_Ashraf%20et%20al.pdf

ASTM E1252-98. (2021). Standard Practice for General Techniques for Obtaining Infrared Spectra for Qualitative Analysis, ASTM International, West Conshohocken, PA. https://doi.org/10.1520/E1252-98R21

ASTM E1508-12. (2019). Standard Guide for Quantitative Analysis by Energy-Dispersive Spectroscopy, ASTM International, West Conshohocken, PA. https://doi.org/10.1520/E1508-12AR19

ASTM C1723-16. (2016). Standard Guide for Examination of Hardened Concrete Using Scanning Electron Microscopy, ASTM International, West Conshohocken, PA. https://doi.org/10.1520/C1723-16

ASTM E168-16. (2016). Standard Practices for General Techniques of Infrared Quantitative Analysis. ASTM International, West Conshohocken, PA. https://doi.org/10.1520/E0168-16

ASTM C143/C143M-20. (2020). Standard Test Method for Slump of Hydraulic-Cement Concrete, ASTM International, West Conshohocken, PA. https://doi.org/10.1520/C0143_C0143M-20

Awoyera, P. O., & Adesina, A. (2020). Plastic wastes to construction products: Status, limitations and future perspective. *Case Studies in Construction Materials, 12*(1), e00330. https://doi.org/10.1016/j.cscm.2020.e00330

Bag, R., Agarwal, A., & Praneeth, R. (2020). Recycling of domestic plastic waste bags as fine aggregate in concrete. *IOP Conference Series: Materials Science and Engineering, 936*, 012009. https://doi.org/10.1088/1757-899X/936/1/012009

Balasegaram, M. (2018) Human Writes: Malaysia is 8th worst in the world for plastic waste. The Star. Viewed at https://www.thestar.com.my/lifestyle/living/2018/03/25/plastic-waste-pollution/

Balea, A., Fuente, E., Monte, M. C., Blanco, A., & Negro, C. (2021). Recycled fibers for sustainable hybrid fiber cement based material: A review. *Materials, 14*(9), 2408. https://doi.org/10.3390/ma14092408

Batayneh, M., Marie, I., & Asi, I. (2007). Use of selected waste materials in concrete mixes. *Waste Management, 1*(1), 1870–1876. https://doi.org/10.1016/j.wasman.2006.07.026

Bethe, S. S., Haque, M. N., & Islam, M. R. (2019). Effects of selected chemical admixtures on physical and mechanical properties of cement mortar. *Progressive Agriculture, 30*(3), 305–310.

Bravard, J.-P., & Gaillot, M. G. E. S. (2013). Geography of sand and gravel mining in the lower Mekong river. *EchoGeo, 26*(1), 1–20. https://doi.org/10.4000/echogeo.13659

BS EN 12390-3. (2019). Testing hardened concrete. Compressive strength of test specimens. British Standard Institution, London, England.

Carabba, L., Manzi, S., & Bignozzi, M. C. (2016). Superplasticizer addition to carbon fly ash geopolymers activated at room temperature. *Materials, 9*(7), 586. https://doi.org/10.3390/ma9070586

Chen, H. L., Nath, T. K., Chong, S., Foo, V., Gibbins, C., & Lechner, A. M. (2021). The plastic waste problem in Malaysia: Management, recycling and disposal of local and global plastic waste. *SN Applied Science, 3*(1), 1–15. https://doi.org/10.1007/s42452-021-04234-y

Choi, Y. W., Moon, D.-J., Chung, J.-S., & Cho, S. K. (2005). Effects of waste PET bottles aggregate on the properties of concrete. *Cement and Concrete Research, 35*(4), 776–781. https://doi.org/10.1016/j.cemconres.2004.05.014

Dash, S. A., Jamaluddin, N., Wan Ibrahim, M. H., Abdullah, A. H., Abdul Awal, A. S. M., Sohu, S., & Ali, N. (2017). Utilization of sugarcane bagasse ash in concrete as partial replacement of cement. *IOP Conference Series: Materials Science and Engineering, 271*(1), 1–10. https://doi.org/10.1088/1757-899X/271/1/012001

Elango, A., & Ashok Kumar, A. (2018). Study on partial replacement of plastic waste as fine aggregate in concrete. *Technical Research Organization India, 5*(5), 1–9. http://troindia.in/journal/ijcesr/vol5iss5/1-9.pdf

Ferreira, L., Brito, J. D., & Saikia, N. (2012). Influence of curing conditions on the mechanical performance of concrete containing recycled plastic aggregate. *Construction and Building, 36*(1), 196–204. https://doi.org/10.1016/j.conbuildmat.2012.02.098

Ferronato, N., & Torretta, V. (2019). Waste mismanagement in developing countries: A review of global issues. *International Research and Public Health, 16*(6), 1060. https://doi.org/10.3390/ijerph16061060

Frigione, M. (2010). Recycling of PET bottles as fine aggregate in concrete. *Waste Management, 30*(6), 1101–1106. https://doi.org/10.1016/j.wasman.2010.01.030

Gavriletea, M. D. (2017). Environmental impacts of sand exploitation. Analysis of sand market. *Sustainability, 9*(7), 2017. https://doi.org/10.3390/su9071118

Gondo, T., Mathada, H., & Amponsah-Dacosta, F. (2019). Regulatory and policy implications of sand mining along shallow waters of Njelele River in South Africa. *Journal of Disaster Risk Studies, 11*(3), 1–13. https://doi.org/10.4102/jamba.v11i3.727

Hama, S. M. (2021). Evalutions of strengths, impact and energy capacity of two-way concrete slabs incorprating waste plastic. *Journal of King Saud University—Engineering Sciences, 33*(5), 337–345. https://doi.org/10.1016/j.jksues.2020.09.007

IS 10262. (2009). Concrete mix proportioning—Guidelines. Indian Standards (pp. 1–14).

Ismail, Z. Z., & Al-Hashmi, E. A. (2008). Use of waste plastic in concrete mixture as aggregate replacement. *Waste Management, 28*(11), 2041–2047. https://doi.org/10.1016/j.wasman.2007.08.023

Jereme, I. A., Siwar, C., & Alam, M. M. (2015). Waste recycling in Malaysia: Transition from developing to developed country. *Indian Journal of Education and Information Management, 4*(1), 1–14.

Kniivila, M. (2007). Industrial development and economic growth: Implications for poverty reduction and income inequality. In J. A. Ocampo (Ed.), *Industrial Development for the 21st Century: Sustainable Development Perspectives* (pp. 295–332). UN Department of Economic and Social Affairs, Division for Sustainable Development. https://www.un.org/esa/sustdev/publications/industrial_development/3_1.pdf

Koting, S., Mahmud, H., & Karim, M. R. (2007). Influence of superplasticizer type and dosage on the workability and strength of cementitious grout for semi-flexible pavement application. *Journal of the Eastern Asia Society for Transportation Studies, 7*(1), 1–7. https://doi.org/10.11175/easts.7.2156

Kou, S., Lee, G., Poon, C., & Lai, W. (2009). Properties of lightweight aggregate concrete prepared with PVC granules derived from scraped PVC pipes. *Waste Management, 29*(2), 621–628. https://doi.org/10.1016/j.wasman.2008.06.014

Lee, Z. H., Paul, S. C., Kong, S. Y., Susilawati, S., & Yang, X. (2019). (2019) Modification of waste aggregate PET for improving the concrete properties. *Advances in Civil Engineering, 1*, 1–10. https://doi.org/10.1155/2019/6942052

McAllister, J. (2015). Factors influencing solid-waste management in the developing world (pp. 1–86). Master Thesis. Utah State University. https://digitalcommons.usu.edu/cgi/viewcontent.cgi?article=1537&context=gradreports

Musbah, M. G., Allam, A. M. A., Saleh, H. A., & Ateeg, I. M. (2019). Effects of superplasticizing admixtures on the compressive strength of concrete. *Universal Journal of Engineering Science, 7*(2), 39–45. https://www.hrpub.org/download/20190430/UJES3-14613063.pdf

Mustafa, M. A., Hanafi, I., Mahmoud, R., & Tayeh, B. A. (2019). Effect of partial replacement of sand by plastic waste on impact resistance of concrete: Experiment and simulation. *Structures, 20*(1), 519–526. https://doi.org/10.1016/j.istruc.2019.06.008

Naik, T. R., Singh, S. S., Huber, C. O., & Brodersen, B. S. (1996). Use of post-consumer waste plastics in cement-based composites. *Cement and Concrete Research, 26*(10), 1489–1492. https://doi.org/10.1016/0008-8846(96)00135-4

Nursyamsi, N., & Adil, K. (2021). The effect of shredded plastic waste PET and HDPE substitution on concrete characteristics. *IOP Conference Series: Materials Science and Engineering, 1122*(1), 012014. https://doi.org/10.1088/1757-899X/1122/1/012014

Olowofoyeku, A. M., Ofuyatan, O. M., Oluwafemi, J., Ajao, A., & David, O. (2019). Effect of superplasticizer on workability and properties of self-compacting concrete. *International Conference on Engineering for Sustainable World, 1378*(1), 042088. https://doi.org/10.1088/1742-6596/1378/4/042088

Patanaik, B. M., & Rath, M. K. (2018). Increase in compressive strength of concrete by addition of super plasticizer. *International Journal of Advanced Technology and Engineering Research, 1*(1), 26–30. http://www.ijater.com/files/NCRTSTM_01_06.pdf

Prusty, J. K., Patro, S. K., & Basarkar, S. S. (2016). Concrete using agro-waste as fine aggregate for sustainable built environment—A review. *International Journal of Sustainable Built Environment, 5*(2), 312–333. https://doi.org/10.1016/j.ijsbe.2016.06.003

Rai, B., Rushad, S. T., Kr, B., & Duggal, S. K. (2012). (2012) Study of waste plastic mix concrete with plasticizer. *International Scholarly Research Notices, 1*, 1–5. https://doi.org/10.5402/2012/469272

Sahmaran, M., Bilici, Z., Ozbay, E., Erdem, T. K., Yucel, H. E., & Lachemi, M. (2013). Improving the workability and rheological properties of engineered cementitious composites using factorial experimental design. *Composites Part B: Engineering, 45*(1), 356–368. https://doi.org/10.1016/j.compositesb.2012.08.015

Saikia, N., & De Brito, J. (2013). Waste polyethylene terephthalate as an aggregate in concrete. *Materiasl Research, 16*(2), 1–10. https://doi.org/10.1590/S1516-14392013005000017

Salem, M., Alsadey, S., & Johari, M. (2016). Effect of superplasticizer dosage on workability and strength characteristics of concrete. *IOSR Journal of Mechanical and Civil Engineering, 13*(4), 153–158. http://iosrjournals.org/iosr-jmce/papers/vol13-issue4/Version-7/T130407153158.pdf

Silva, R., De Brito, J., & Saikia, N. (2013). Influence of curing conditions on the durability-related performance of concrete made with selected plastic waste aggregates. *Cement and Concrete Composites, 35*(1), 23–31. https://doi.org/10.1016/j.cemconcomp.2012.08.017

Stefan, M., Salvador, V., & David, S. (2018). Validation of artificial defects for Non-destructive testing measurements on a reference structure. *MATEC Web of Conferences, 199*(1), 06006. https://doi.org/10.1051/matecconf/201819906006

Thompson, R. C., Moore, C. J., Sal, F. S. V., & Swan, S. H. (2009). Plastics, the environment and human health: Current consensus and future trends. *Philosophical Transactions of the Royal Society B: Biological Science, 364*(1526), 1–20. https://doi.org/10.1098/rstb.2009.0053

Thoo, F. V. W., Zainuddin, N., Matori, K. A., & Abdullah, S. A. (2013). Studies on the potential of waste soda lime silica glass in glass ionomer cement production. *Advances in Materials Science and Engineering, 2013*(1), 1–6. https://doi.org/10.1155/2013/395012

Thorneycroft, J., Orr, J., Savoikar, P., & Ball, R. J. (2018). Performance of structural concrete with recycled plastic waste as a partial replacement for sand. *Construction and Building Materials, 161*(1), 63–69. https://doi.org/10.1016/j.conbuildmat.2017.11.127

Turku, I., Karki, T., Rinne, K., & Puurtinen, A. (2016). Characterization of plastic blends made from mixed plastics waste of different sources. *Waste Management & Research: the Journal for a Sustainable Circular Economy, 35*(2), 200–206. https://doi.org/10.1177/0734242X16678066

Umasabor, R. I., & Daniel, S. C. (2020). The effect of using polyethylene terephthalate as an additive on the flexural and compressive strength of concrete. *Heliyon, 6*(8), e04700. https://doi.org/10.1016/j.heliyon.2020.e04700

Xun, W., Wu, C., Leng, X., Li, J., Xin, D., & Li, Y. (2020). Effect of functional superplasticizers on concrete strength and pore structure. *Applied Sciences, 10*(10), 3496. https://doi.org/10.3390/app10103496

Yurdakul, E., Taylor, P. C., Ceylan, H., & Bektas, F. (2014). Effect of water-to-binder ratio, air content, and type of cementitious materials on fresh and hardened properties of binary and ternary blended concrete. *Journal of Materials in Civil Engineering, 26*(6), 1–20. https://doi.org/10.1061/(ASCE)MT.1943-5533.0000900

Zulkernain, N. H., Gani, P., Chuan, N. C., & Uvarajan, T. (2021). Utilisation of plastic waste as aggregate in construction materials: A review. *Construction and Building Materials, 296*(1), 123669. https://doi.org/10.1016/j.conbuildmat.2021.123669

Ceramic Tiles Waste as Coarse Aggregate Filler Replacement in Concrete

Chin Mei Yun, Md Rezaur Rahman, Durul Huda, Kuok King Kuok, Amelia Chai Pei Sze, Rudison Anak Sering, and Muhammad Khusairy Bin Bakri

Abstract This research looked at varying percentages of natural coarse aggregate partially substituted with Coarse Ceramic Aggregate (CCA) in the same mixture design as normal-weight concrete. CCA can be obtained from construction waste due to the end of the life cycle of a building after demolition. This research looks at the behaviour of various percentages of partial replacement with CCA in terms of compressive strength and workability in each concrete mixture cast. According to the findings, the ideal proportion of partially replaced CCA is approximately 10% to 30%, resulting in a considerably greater compressive strength output. Meanwhile, partial replacement at 40% revealed that compressive strength was about the same as normal-weight concrete. Furthermore, of all the percentages, the 50% partially replacement batch has the lowest compressive strength and is significantly lighter than normal-weight concrete. In a word, CCA is feasible to be adopted as construction material in terms of application to partially replace regular coarse aggregate in concrete casting due to its compressive strength characteristics and workability behaviour.

Keywords Ceramic · Coarse aggregate · Concrete · Partial replacement · Waste · Tiles

1 Introduction

Ceramic tiling ware is an essential component of construction materials in buildings and homes nowadays (Nara et al., 2014; Subedi, 2013). The importance of having

C. M. Yun · K. K. Kuok · A. C. P. Sze · R. A. Sering
Faculty of Engineering, Computing and Science, Swinburne University of Technology Sarawak Campus, Jalan Simpang Tiga, 93400 Kuching, Sarawak, Malaysia

M. R. Rahman (✉) · M. K. B. Bakri
Faculty of Engineering, Universiti Malaysia Sarawak, Jalan Datuk Mohammad Musa, 94300 Kota Samarahan, Sarawak, Malaysia
e-mail: rmrezaur@unimas.my

D. Huda
Department of Mechanical Engineering and Product Design Engineering, Swinburne University of Technology, Hawthorn, VIC 3122, Australia

M. R. Rahman et al. (eds.), *Waste Materials in Advanced Sustainable Concrete*, Engineering Materials, https://doi.org/10.1007/978-3-030-98812-8_10

or employing ceramic materials in any building or residence has become one of the trends that today's market must quickly adapt to (Ali & Al-Kodmany, 2012). To meet the market's need for ceramic tiling ware, manufacturers must create a considerable volume of the product. Aside from that, a slew of new manufacturing companies is vying for market share with sophisticated designs and a wide range of colour options for potential customers. These circumstances resulted in overproduction since a substantial portion of the ceramic tile produced was unable to be purchased by consumers owing to the design or colour choices that were out of step with current market demand.

Furthermore, obsolete and outdated ceramic tiling may be challenging to sell and may wind up stacking up in the junkyard before being turned into ceramic trash (Abu Bakar, 2020). Furthermore, ceramic waste material might be generated on the job site due to ceramic tiling that was not up to specification and was rejected by the contractor, as well as ceramic tiling that was damaged during handling and shipping the material (Casagrande et al., 2002; Hung & Kamaludin, 2017; Vieira, & Monteiro, 2009). Ceramic waste accounts for about one-third of all building trash (Esin & Cosgun, 2007). Ceramic waste is difficult to recycle into fresh ceramic tiling or any other kind of new ceramic due to its unique characteristics (El-Kattan et al., 2020). Thus, it is critical to resolve ceramic waste and transform it into something useful in terms of construction materials by reusing or recycling it into something that matches the strength and durability of ceramic waste qualities, mainly if it is to be used as partly coarse aggregate in concrete (Huang et al., 2009).

According to Senthamarai and Manoharan (2005), concrete producers must employ waste hazardous industrial materials to create a greener environment. Currently, over 30% of trash is not recycled, with most of it coming from the ceramics sector (Needhidasan et al., 2014). There were experimentation and study on the possibility of replacing traditional crushed natural coarse aggregate with ceramic industrial waste (Ikponmwosa & Ehikhuenmen, 2017; Senthamarai & Manoharan, 2005; Tadesse, 2020). In terms of compressive, splitting, tensile, and flexural strength, as well as modulus of elasticity, experiments were conducted comparing conventional concrete produced with natural coarse aggregate and ceramic waste coarse aggregate (Senthamarai & Manoharan, 2005). The results of the tests on both concretes were compared, and it was found that ceramic waste coarse aggregated concrete had better workability and strength than conventional concrete, prepared using natural coarse aggregate (Awoyera et al., 2018; Ikponmwosa & Ehikhuenmen, 2017; Senthamarai and Manoharan, 2005). However, this research study has a drawback in that it only looks at ceramic electrical insulator industrial waste. The factor contributing to the ineffectiveness of using this material is that it is considered hazardous to handle, even though it can be recycled into ceramic waste with proper extraction and separation by workforce or machinery (Troschinetz, 2005). Getting it in terms of the ceramic waste product would slow down the processing line. Apart from that, the amount of ceramic waste source industrial electrical insulator may not be available in large quantities due to the low frequency of broken or defective material among the subjected materials, making it difficult to produce large quantities of ceramic waste if it is to be used as coarse aggregate in concrete (Czajczynska et al., 2017).

With the country's fast modernization, the need for concrete with natural coarse aggregate has risen rapidly throughout the year (Collivignarelli et al., 2020). The demand for concrete with natural coarse aggregate is rising in the building industry because of the new development (Karim et al., 2016; Xiao et al., 2018). This material was used to make the foundation of a building, particularly a high-rise structure that required a large volume of concrete with natural coarse aggregate (Poulos, 2016). The amount or ratio of the natural coarse aggregate stone cast with concrete mix increases as the demand for concrete strength increases (Alsayed & Amjad, 1996). The proportion of aggregate in the concrete determines the concrete's strength (Kang & Weibin, 2018; Kozul & Darwin, 1997). As a result, concrete with aggregate is widely used to construct megastructures, notably tall and large structures (Meng et al., 2021). The issue arises when a limited supply of natural coarse aggregate stone varies based on the region's geology. If natural resources, particularly coarse aggregate stone to be mixed with concrete, are in short supply, these might stymie the project's progress (Makul et al., 2021; Nakhi & Alhumoud, 2019). As a result, a novel and cost-effective solution must be implemented to eliminate or minimize the reliance on natural coarse aggregate stone by partially replacing coarse aggregate in concrete with ceramic waste (Torkittikul & Chaipaich, 2010). When partly aggregated concrete is replaced with ceramic waste tile, the strength of the concrete is improved. It reduces the demand for natural aggregate and increases the demand for recycling ceramic waste for concrete. It also would produce a better result in concrete strength and the sustainability of having ceramic waste that could be utilized for a greater purpose.

2 Methodology

2.1 *Materials*

Waste tiles were utilized and turned into CCA in this experiment. It was obtained locally from construction waste as a result of building demolition. River sand as fine and gravel as coarse aggregates are obtained from local suppliers through local hardware shops. Ordinary Portland Cement Grade 25 were obtained from Cahaya Mata Sarawak (CMS).

2.2 *Sample Mix Proportion, Casting and Curing*

The concrete mixture contains 10%, 20%, 30%, 40%, and 50% ceramic waste, with one standard coarse aggregate concrete serving as an experimental reference. According to AS 1726 (2017), the size of coarse grain that is enough to be used is between 20 and 2 mm. Thus all the aggregates need to sieve to get the correct size of coarse aggregate. This process needs to be done in the sand and average crushed

Table 1 Sample mix proportion

Concrete mix	Coarse aggregate (%)		Cement (kg/m^3)	Sand (kg/m^3)	Gravel (kg/m^3)	Crushed tiles (CCA) (kg/m^3)	Water (kg/m^3)
	Gravel (%)	Crushed tiles (CCA) (%)					
NC	100	0	380	640	1340.57	0	170
CCA10%	90	10	380	640	1208	120	174.5
CCA20%	80	20	380	640	1074	241	174.5
CCA30%	70	30	380	640	939	359	174.5
CCA40%	60	40	380	640	805	481	174.5
CCA50%	50	50	380	640	671	603	174.5

coarse aggregate to minimize the chance of inconsistency during casting the cube concrete. According to AS 1726 (2017), the ceramic waste tile must be physically smashed with a sledgehammer until the waste is broken down into smaller pieces that meet the coarse grain size. According to the recommended ratio, the required portion of ceramic waste aggregate was combined with gravel, river sand, portable water, and cement Grade-25. Because the water content was a continuous parameter in this experiment, it must be measured. More portable water can be added to the concrete to make it workable before cube casting. The cube concrete was cast in multiples of an average of three cubes for each curing duration of 7, 14 and 28 days, according to IS 10262 (2009) standard requirements. The same techniques were performed for regular gravel as coarse aggregate concrete mixes with consistent water measurements. Table 1 shows the sample mix proportion.

2.3 Workability Test

A slump test was carried out to check for the workability of fresh concrete before the concrete mixture was placed in the moulds. The slump test was carried out under IS 1199 (1959), concrete sampling and analysis procedures.

2.4 Compressive Test

Compressive strength test was conducted using Unit Test Autocon 2000 (Promat (HK) Ltd., Hong Kong) under BS EN 12390-3 (2019). Loading applied to the specimens at the rate of 2.4 ± 0.2 kN/s until failure. The results were averaged and reported for each concrete mix.

2.5 *Scanning Electron Microscope (SEM) and Energy Dispersive X-Ray/Spectroscopy (EDS/EDX)*

SEM and energy dispersive X-ray/spectroscopy (EDX/EDS) were performed in accordance with ASTM C1723-16 (2016) and ASTM E1508-12 (2019) standards, respectively. The samples were examined at magnifications of 100×, 250×, 500× , and 10,000×. The elemental composition percentages of the samples are scanned and analyzed by automated software. The EDX/EDS was performed many times on each sample at various stages, with the most representative results chosen. For the materials' SEM and EDX/EDS examinations, Hitachi Ltd., Tokyo, Japan, utilized a Hitachi TM4000Plus Tabletop Microscope with a Quantax75TM Series Energy Dispersive X-Ray Spectrometer (Hitachi Ltd., Tokyo, Japan).

2.6 *Fourier Transform Infrared Spectroscopy*

Fourier-transform infrared spectroscopy (IRAffinity-1, Shimadzu Corporation, Kyoto, Japan) was used for the FTIR analysis of the samples. Fourier-transform infrared spectroscopy was conducted according to the ASTM E168-16 (2016) and ASTM E1252-98 (2021) standards for qualitative and quantitative analysis. The spectrum scanning was conducted in the wavenumber range of 4000–400 cm^{-1} for each sample. Fourier-transform infrared spectroscopy utilized the samples' infrared spectrum transmittance and absorption to develop a unique molecular fingerprint spectrum. The test was repeated numerous times for each sample, and the most representative results were selected.

3 Results and Discussions

3.1 *Workability Properties*

Figure 1 shows the slump test. The water proportion ratio in this concrete mixture was kept as constant as feasible for all batches of concrete to differentiate fresh concrete workability for each of the various percentages of CCA and NC mix. However, if the mixture was unable to mix appropriately or pass the slump test, water was adjusted as needed to enable the concrete mix to be thoroughly mixed, allowing it to pass slump testing and decreasing the risk of the concrete mix forming honeycomb during curing due to a lack of water (Hoang & Pham, 2016).

Figure 2 shows that the slump test for NC is 25.6 mm, which may be classed as optimum workability according to IS 1199 (1959), indicating that the degree of workability for the mass concrete category is low. Therefore, the workability obtained for CCA10%, CCA20%, CCA 30%, and CCA 40% is 25.8 mm, 36.7 mm, 52.3 mm,

Fig. 1 Slump test

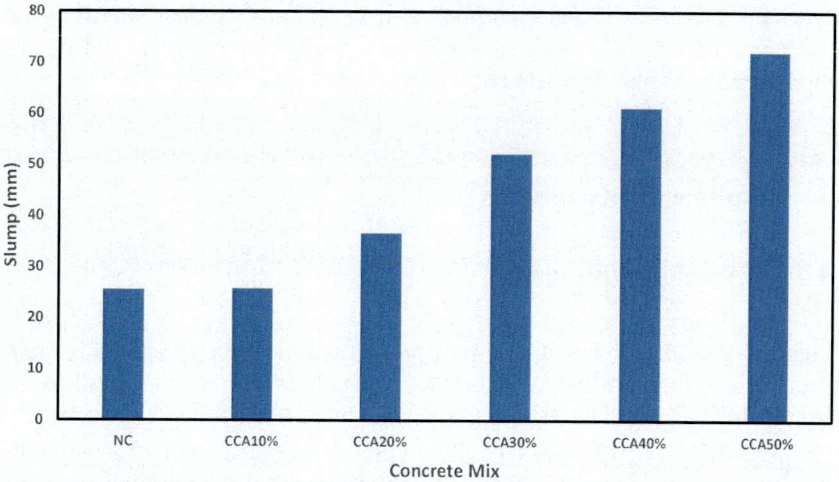

Fig. 2 Slump test of mix concrete

and 61.3 mm, respectively. The greatest slump test result was reported by CCA50%, which indicated a hang of 72.3 mm. The increase in slump test results owing to variations in CCA % is related to the ceramic tiles that were partially replaced in the concrete mix and act as a minor water absorption medium (Correia et al., 2004). It is due to the ability of porcelain in ceramic tiles to absorb a tiny quantity of water, which contributes to water adjustment, which influences the slump test conclusion (Medina et al., 2013). The more ceramic concrete partially replaced in mixed concrete, the less workability in the fresh concrete mixture. However, the water adjustment based on the water ratio percentage was unaffected because a minimal amount of water was added to the concrete mixture to form properly fresh concrete.

3.2 Compressive Strength Properties

Figure 3 shows the average compressive strength of concrete age with the normal and ceramic mixture cubes displayed on a graph. Each batch averaged over five concrete cubes in the compressive strength test. Each batch was cast using the same process with standard weight concrete, with the amount of coarse ceramic aggregate partially substituted with normal coarse aggregate varying. Having the same methodology, design, and mixing percentage in the experiment would provide a more precise conclusion and a more trustworthy collection of data to compare.

CCA50% cubes had lower compressive strength than NC cubes at any age of curing, according to the overall results. When comparing the CCA10%, CCA20,

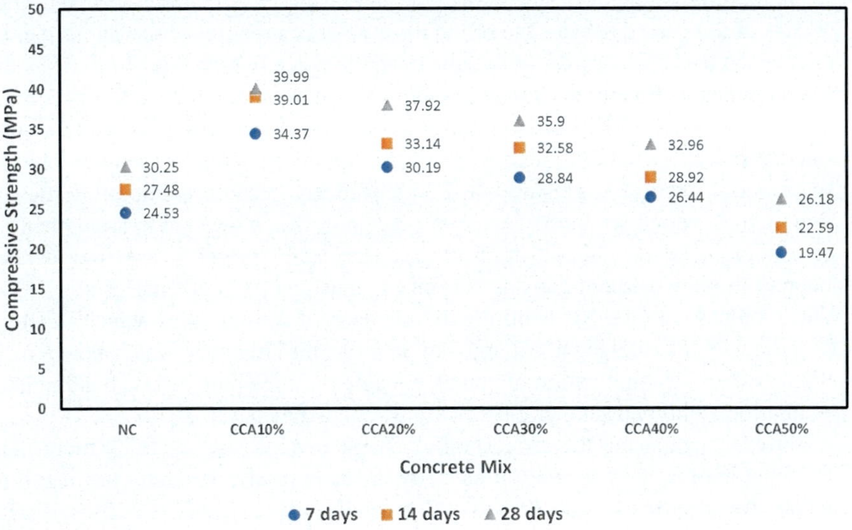

Fig. 3 Compressive strength of concrete age with normal mix and ceramic mix cubes

Fig. 4 NC and CCA10% cube concrete post-testing conditions

CCA30%, and CCA40% cubes to the NC cubes at all ages, the CCA10%, CCA20, CCA30%, and CCA40% cubes exhibited a considerably better average compressive strength test result. Meanwhile, compared to the NC, which served as a control combination for these studies, CCA40% produced approximately equal or nearly comparable average compressive strength test results (Figs. 4 and 5).

The optimal strength of each concrete cube batch was entirely achieved after 28 days of the curing phase. One of the most notable strengths observations can be observed in the results of the 10% cube concrete batch, which marked itself as the most significant strength level in terms of compressive strength, which was considerably more remarkable when compared to the NC control cube other CCA cube concrete batch. It could be interpreted as the condition or microstructure between the ceramic. Average aggregate was bonding nicely because of the properties of ceramic tiles, which are mostly made of porcelain, which may aid in strengthening the bonding between ceramic and average aggregate by preserving an interlocking concrete mixture without causing the mix to fall apart (Andrzejuk et al., 2018). When compared to the NC control cube, the lack of ceramic tiles, which function as a porcelain interlocking agent and may improve the strength of the concrete mix, indicates that the combination of concrete inside the cube may be less bonding and not interlock with one another (Ipomwosa & Ehikhenmen, 2017).

However, increasing the proportion of coarse ceramic aggregate by more than 10% may approach its maximum capacity, as compressive strength has begun to decline. It is also seen in the 20% to 40% partially substituted CCA in the concrete batch. The duration of interlocking between the coarse aggregate in the cube may be reduced in this stage because the concrete mix is less effective in generating better

Fig. 5 All the batch of cube concrete post-testing conditions

bonding that protects the cube's integrity during the compression test (Yehia et al., 2020). Because its strength in terms of compression is the same as the control cube of the NC concrete batch, the resulting outcome defined by the 40% batch of concrete mix is classified as the threshold point for the capability CCA concrete cube. It marked the end of the concrete mix employing coarse ceramic aggregate as a partial replacement, as the compressive strength was insufficient for industrial use below the NC strength (Babatola & Arum, 2020; Prakash & Rao, 2016).

3.3 Morphological Properties

Figure 6a–d shows the SEM images for CCA10% (28 days) at a magnification of 100×, 250×, 500×, and 1000×, respectively. While Fig. 7a–d shows the SEM images for CCA40% (28 days). Based on Fig. 6, it is noted that there is a micro crack observed when magnified 1000x (as shown in Fig. 6a). It was due to micro void, which is common for concrete to dry. In Fig. 7, it is obvious that the tiles waste in CCA40% to be observed more than the those in CCA10%. It also shows a few areas where tiles are bonded together, while some are not properly bonded. It was due to two different surfaces of the tile. The smooth tile surface tends to have limited grip, while the rough surface bonds with the paste, enhancing the cement structure.

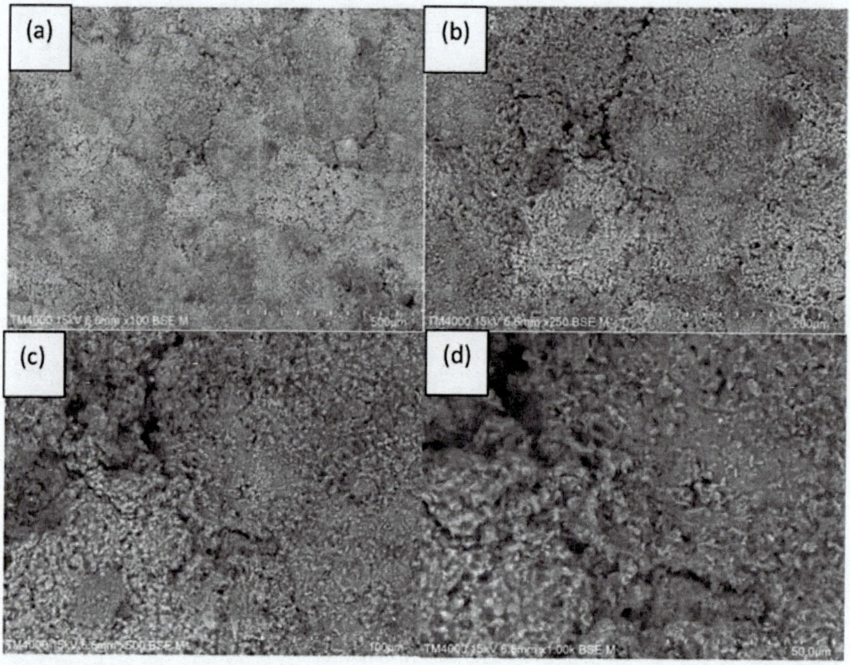

Fig. 6 SEM images for CCA10% (28 days) at magnification of **a** 100×, **b** 250×, **c** 500×, and **d** 1000×

In contrast, the smooth surface might initiate failure, such as small cracks or stress propagation in the concrete (Zegardlo et al., 2018).

3.4 Elemental Properties

Table 2 and Fig. 8 shows the EDS/EDX plot and element contents for CCA10% (28 days). While Table 3 and Fig. 9 shows the EDS/EDX plot and element contents for CCA40% (28 days). The elements present in the sample for Tables 2 and 3 are similar, which is oxygen (O), calcium (Ca), carbon (C), silicon (Si), aluminium (Al), iron (Fe), and sulfur (S). A few amounts of magnesium (Mg) and potassium (K) found in Table 2 extra elements obtained in the tiles waste are expected. Both Tables shows that oxygen mass percentage has the highest content, followed by calcium. It also shows that chemical reactions happened during the mixed paste mixture and hydraulic reactivity (Tabak et al., 2012). At a mature age, the concrete exhibit's higher strength.

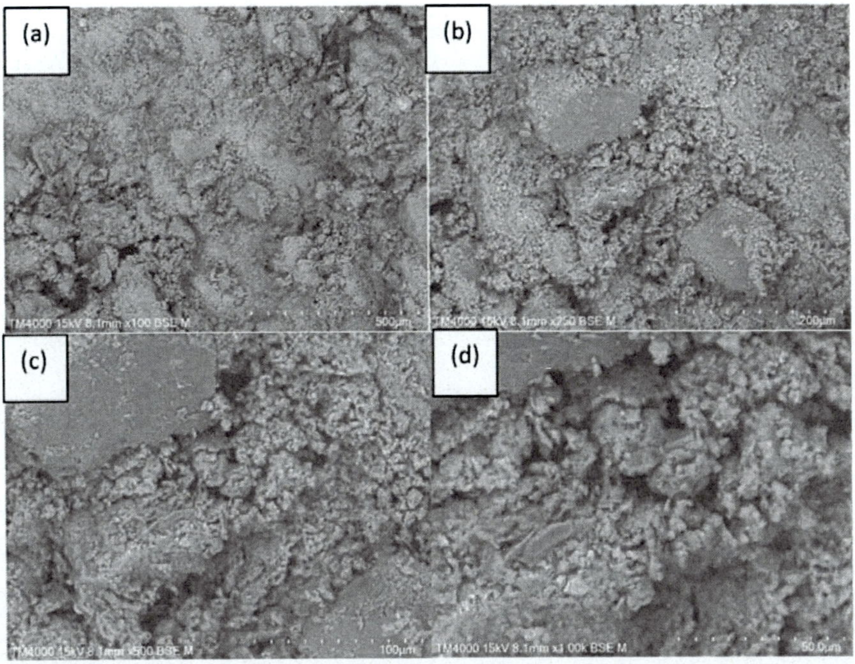

Fig. 7 SEM images for CCA10% (28 days) at magnification of **a** 100×, **b** 250×, **c** 500×, and **d** 1000×

Table 2 Element composition for CCA10% (28 days)

Element	Mass normal (%)	Atom (%)
O	43.56	53.67
Ca	21.60	10.62
C	12.52	20.55
Si	10.59	7.43
Al	5.27	3.85
Mg	2.61	2.12
Fe	1.81	0.64
K	1.20	0.60
S	0.84	0.52
Sum	100.00	100.00

3.5 *Infrared Spectral Properties*

Figures 10 and 11 shows the FTIR images for CCA10% and CCA40% at 28 days, respectively. The small broad peak at the 3200–3500 cm^{-1} band was due to the O-H bond due to the hydraulic of water bond and moisture. It appeared on both

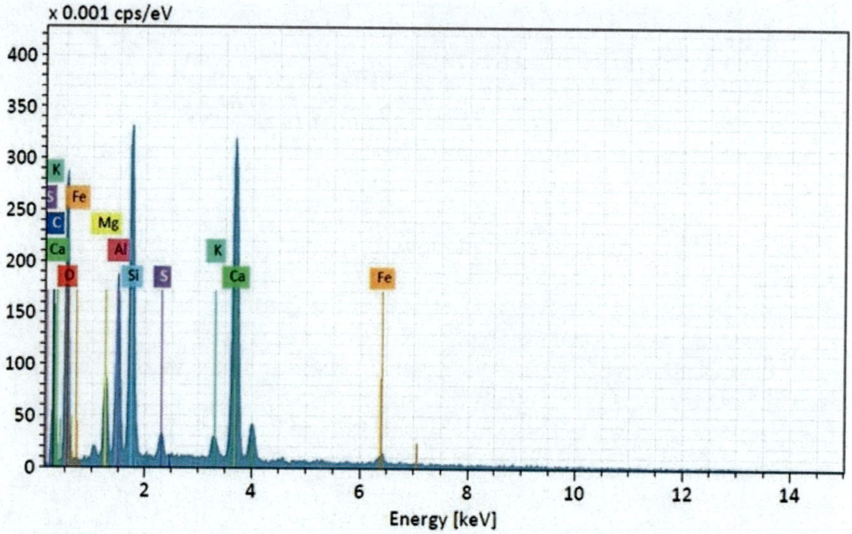

Fig. 8 EDS/EDX graph for CCA10% (28 days)

Table 3 Element composition for CCA40% (28 days)

Element	Mass normal (%)	Atom (%)
O	54.07	69.39
Ca	29.49	15.11
Si	7.78	5.69
C	4.10	7.02
Al	2.20	1.67
Fe	1.42	0.52
S	0.94	0.60
Sum	100.00	100.00

figures but lowered peak size for CCA10%. It is responsible for the microcrack in the concrete. An analysis of the peak at 2555.68 cm^{-1}, 2453.45 cm^{-1}, 2181.49 cm^{-1}, and 2005.97 cm^{-1} was due to various FTIR spectra C-H bending and stretching. The most common wavelength is 582.22 cm^{-1} was due to symmetric and asymmetric valence and deformation vibration, respectively (Jorda et al., 2015).

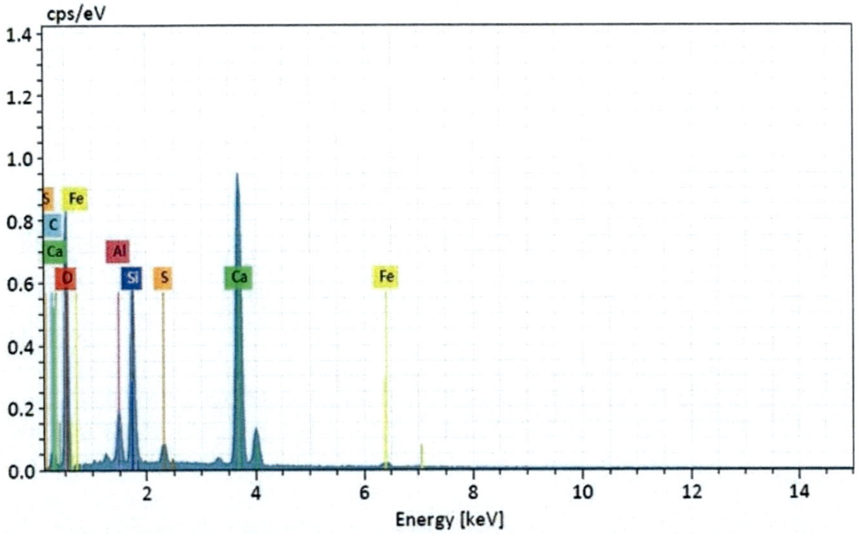

Fig. 9 EDS/EDX graph for CCA40% (28 days)

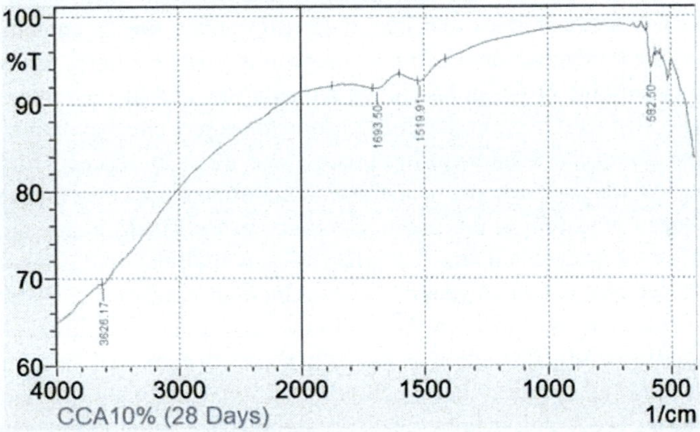

Fig. 10 FTIR for CCA10% (28 days)

4 Application and Benefit of Course Ceramic Tiles Waste as Replacement in Concrete

In America, more than 40% of ceramic waste will be recycled, with the remainder being converted into ceramic powder and sent to other firms that may utilize it (Daigo et al., 2018). According to statistics published by the British Ceramic Research Association, the recycling rate of most ceramic companies has surpassed 40% (Ncube

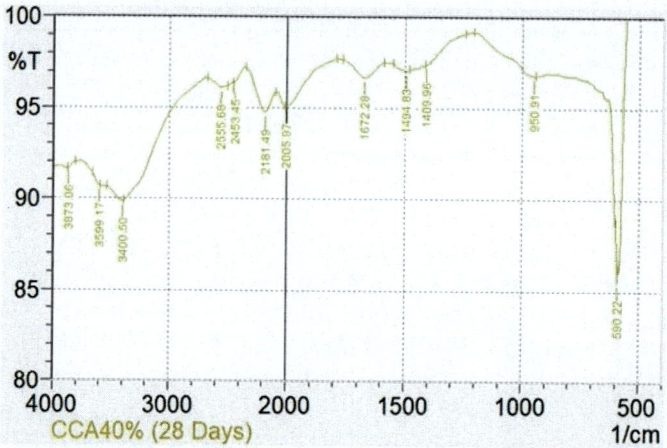

Fig. 11 FTIR for CCA40% (28 days)

et al., 2021). The primary method of re-utilization is to reprocess discarded porcelain and use it to manufacture fresh porcelain. If the product is white ceramic tile, we cannot utilize red or other dark-coloured ceramic wastes, and companies that process waste ceramics and materials will contact tile manufacturers at any moment. This system for processing and recycling waste ceramics has received the approval of ceramic professionals, environmentalists, and the government, as well as multiple awards (Daigo et al., 2018). Companies pay close attention to ceramic waste throughout the manufacturing process and insist on reprocessing and recycling. Due to a lack of resources and a high environmental consciousness, several nations began researching the use of ceramic wastes 10–20 years ago and have reasonably sophisticated dealing technology. Many multinational firms have started to employ ceramic wastes to generate porous thermal insulation construction materials by combining polished wastes with other materials. The utilization rate has almost approached 100% (Amoros et al., 2007; Zanelli et al., 2021). In recent years, most ceramics-building firms have included a belt-type rotating mill device into their manufacturing processes, reaping relatively significant economic and environmental advantages.

Senthamarai and Manoharan (2005) developed the tests on waste ceramic pieces, totally replacing the standard coarse aggregate in 2003, which sets the maximum size of coarse aggregate at 20. The experimental results show that, while the alternative experimental group has an advantage in low tension and compression ratio and workability, the alternative ones have lower compressive strength, splitting tensile strength, and flexural strength than the traditional raw materials of concrete by 3.8%, 18.2%, and 6%, respectively. Similarly, Brito et al. (2005) conducted more extensive tests in researching ceramic trash alternatives to conventional coarse aggregate, utilizing ceramic pieces of 1/3, 2/3, and entirely traditional coarse aggregate. The experimental results revealed that: the wear of the alternative group increased, but the compressive and flexural strength of the alternative concrete block set decreased as the replacement ratio increased. The compressive strength reduction value was more significant than the flexural strength reduction value.

Instead of fine aggregate, in the research of ceramic waste, pieces of ceramic waste are always reprocessed first and then polished to form fine aggregate as the particle with a maximum size of 4 mm. Binici (2007) and Lopez et al. (2007) conducted mechanics experiments with different substitution rates of 40%, 60% and 10% 50%, respectively. Their experimental results showed: In wear resistance, workability, and resistance to chloride ion penetration, the concrete group of substitution is better than traditional concrete, and with the increase of replacement ratio, the compressive strength increased and was higher than that of conventional concrete.

5 Conclusion

Based on the experiment results, a conclusion can be formed based on the performance study that contributed to verifying and developing information linked to the partial replacement of coarse ceramic aggregate in regular concrete. From the standpoint of workability, the coarse ceramic aggregate would degrade in proportion to the extra percentage of partial replacement toward the regular coarse aggregate. In other words, the lower the workability of the concrete mixture, the more CCA is partially replaced in normal-weight concrete. Aside from that, another conclusion that can be drawn from the experiment is that the CCA10% concrete batch had the highest compressive strength. The concrete combination might be used in industrial applications requiring high strength concrete, particularly under compression, with such a fantastic result. Aside from that, CCA20% and CCA30% can be considered positive results from this experiment because both of these batches managed to suppress the control cube of NC in terms of compressive strength test but capped below the strength of CCA10%due to the increase in percentage in partial replacement of CCA, which may reduce small portion bonding inside the concrete mixture. The CCA20% and CCA30% concrete batches could produce good compressive strength and performance, especially in construction industry applications. The CCA40% showed virtually acceptable compressive strength with the NC concrete batch. Although this batch did not perform as well as others with a lower CCA percentage than the CCA40% batch, it produced encouraging results since it was more likely to match its compressive strength to the NC control cube. This CCA40% batch of concrete could also be recommended if it were used in the construction industry, which may be inquiring a large number of ceramic tiles waste due to demolished work or renovation work on-site, and it could potentially save money by partially replacing natural coarse aggregate with coarse ceramic aggregates, particularly from waste tiles.

Acknowledgements The authors would like to thank Swinburne University of Technology Sarawak Campus and Universiti Malaysia Sarawak (UNIMAS) for the collaboration.

References

Abu Bakar, M. A. B. (2020). A review of ceramic tiles waste as an alternative construction material (pp.1–95). Degree Thesis, UTHM.

Ali, M. M., & Al-Kodmany, K. (2012). Tall buildings and urban habitat of the 21st century: A global perspective. *Buildings, 2*(4), 384–423. https://doi.org/10.3390/buildings2040384

Alsayed, S. H., & Amjad, A. (1996). Strength, water absorption and porosity of concrete incorporating natural and crushed aggregate. *Journal of King Saud University—Engineering Sciences, 8*(1), 109–119. https://doi.org/10.1016/S1018-3639(18)30642-1

Amoros, J. L., Orts, M. J., Garcia-Ten, J., Gozalbo, A., & Sanchez, E. (2007). Effect of the green porous texture on porcelain tile properties. *Journal of the European Ceramic Society, 27*(5), 2295–2301. https://doi.org/10.1016/j.jeurceramsoc.2006.07.005

Andrzejuk, W., Barnat-Hunek, D., Siddique, R., Zegardlo, B., & Lagod, G. (2018). Application of recycled ceramic aggregates for the production of mineral-asphalt mixtures. *Materials, 11*(5), 658. https://doi.org/10.3390/ma11050658

AS 1726. (2017). Geotechnical site investigations. Standards Australia (pp.1–10).

ASTM E1252-98. (2021). Standard Practice for General Techniques for Obtaining Infrared Spectra for Qualitative Analysis. ASTM International, West Conshohocken, PA. https://doi.org/10.1520/E1252-98R21

ASTM E1508-12. (2019). Standard Guide for Quantitative Analysis by Energy-Dispersive Spectroscopy. ASTM International, West Conshohocken, PA. https://doi.org/10.1520/E1508-12AR19

ASTM C1723-16. (2016). Standard Guide for Examination of Hardened Concrete Using Scanning Electron Microscopy. ASTM International, West Conshohocken, PA. https://doi.org/10.1520/C1723-16

ASTM E168-16. (2016). Standard Practices for General Techniques of Infrared Quantitative Analysis. ASTM International, West Conshohocken, PA. https://doi.org/10.1520/E0168-16

Awoyera, P. O., Ndambuki, J. M., Akinmusuru, J. O., & Omole, D. O. (2018). Characterization of ceramic waste aggregate concrete. *HRBC Journal, 14*(3), 282–287. https://doi.org/10.1016/j.hbrcj.2016.11.003

Babatola, O., & Arum, C. (2020). Determination of the compressive strength of concrete from binary cement and ternary aggregates. *Open Journal of Civil Engineering, 10*(4), 1–10. https://doi.org/10.4236/ojce.2020.104029

Binici, H. (2007). Effect of crushed ceramic and basaltic pumice as fine aggregates on concrete mortars properties. *Construction and Building Materials, 21*(6), 1191–1197. https://doi.org/10.1016/j.conbuildmat.2006.06.002

Brito, J. D., Pereira, A. S., & Correria, J. R. (2005). Mechanical behaviour of non-structural concrete made with recycled ceramic aggregates. *Cement and Concrete Composites, 27*(4), 429–433. https://doi.org/10.1016/j.cemconcomp.2004.07.005

BS EN 12390-3. (2019). Testing hardened concrete. Compressive strength of test specimens. British Standard Institution, London, England.

Casagrande, M. C., de Oliveira, A. P. N., Hotza, D., da Silva, O. A., & Alexandre, L. R. (2002). Recycling of solid waster in the production of ceramic floor tiles. *Castellon,* 39–46. https://www.qualicer.org/recopilatorio/ponencias/pdfs/0232090e.pdf

Collivignarelli, M. C., Cillari, G., Ricciardi, P., Miino, M. C., Toretta, V., Rada, E. C., & Abba, A. (2020). The production of sustainable concrete with the use of alternative aggregates: A review. *Sustainability, 12*(9), 7903. https://doi.org/10.3390/su12197903

Correia, S. L., Hotza, D., & Segadaes, A. M. (2004). Simultaneous optimization of linear firing shrinkage and water absorption of triaxial ceramic bodies using experiments design. *Ceramics International, 30*(6), 917–922. https://doi.org/10.1016/j.ceramint.2003.10.013

Czajczynska, D., Anguilano, L., Ghazal, H., Kryzynska, R., Reynolds, A. J., Spencer, N., & Jouhara, H. (2017). Potential of pyrolysis processes in the waste management sector. *Thermal Science and Engineering Progress, 3*(1), 171–197. https://doi.org/10.1016/j.tsep.2017.06.003

Daigo, I., Kiyohara, S., Okada, T., Okamoto, D., & Goto, Y. (2018). Element-based optimization of waste ceramic materials and glasses recycling. *Resources, Conservation and Recycling, 133*(1), 375–384. https://doi.org/10.1016/j.resconrec.2017.11.012

El-Kattan, I. M., Abdelzaher, M. A., & Farghali, A. A. (2020). Positive impact of ultra fine-ceramic waste on the physico-mechanical features and microstructure of white cement pastes composites. *Journal of Materials Research and Technoology, 9*(4), 9395–9402. https://doi.org/10.1016/j.jmrt.2020.05.087

Esin, T., & Cosgun, N. (2007). A study conducted to reduce construction waste generation in Turkey. *Building and Environment, 42*(4), 1667–1674. https://doi.org/10.1016/j.buildenv.2006.02.008

Hoang, N.-C., & Pham, A.-D. (2016). Estimating concrete workability based on slump test with least squares support vector regression. *Journal of Construction Engineering, 2016*(1), 1–8. https://doi.org/10.1155/2016/5089683

Huang, B., Dong, Q., & Burdette, E. G. (2009). Laboratory evaluation of incorporating waste ceramic materials into Portland cement and asphaltic concrete. *Construction and Building Materials, 23*(12), 3451–3456. https://doi.org/10.1016/j.conbuildmat.2009.08.024

Hung, F. C., & Kamaludin, N. S. (2017). Professionals' views on material wastage level and causes of construction waste generation in Malaysia. *Malaysian Construction Research Journal, 21*(1), 33–54.

Ikponmwosa, E. E., & Ehikhuenmen, S. O. (2017). The effect of ceramic waste as coarse aggregate on strength properties of concrete. *Nigerian Journal of Technology, 36*(3), 691–696.https://doi.org/10.4314/njt.v36i3.5

IS 10262. (2009). Concrete mix proportioning—Guidelines. Indian Standards, Delhi.

IS 1199. (1959). Methods of sampling and analysis of concrete. Indian Standard, Delhi.

Jorda, J. D., Jordan, M. M., Ibanco-Canete, R., Reyes-Labarta, J. A., Sanchez, A., & Cerdan, M. (2015). Mineralogical analysis of ceramic tiles by FTIR: A quantitative attempt. *Applied Clay Science, 115*(1), 1–8. https://doi.org/10.1016/j.clay.2015.07.005

Kang, M., & Weibin, L. (2018). Effect of the aggregate size on strength properties of recycled aggregate concrete. *Advances in Materials Science and Engineering, 2018*(1), 1–8. https://doi.org/10.1155/2018/2428576

Karim, Y., Khan, Z., Alsoufi, M. S., & Yunus, M. (2016). A review on recycled aggregates for the construction industry. *American Journal of Civil Engineering and Architecture, 4*(1), 27–33. http://pubs.sciepub.com/ajcea/4/1/5

Kozul, R., & Darwin, D. (1997). Effects of aggregate type, size, and content on concrete strength and fracture energy. *Structural Engineering and Engineering Materials, 43*(1), 1–98.

Lopez, V., Llamas, B., Juan, A., Moran, J. M., & Guerra, I. (2007). Eco-efficient concretes: Impact of the use of white ceramic powder on the mechanical properties of concrete. *Biosystems Engineering, 96*(4), 559–564. https://doi.org/10.1016/j.biosystemseng.2007.01.004

Makul, N., Fediuk, R., Amran, M., Zeyad, A. M., Klyuev, S., Chulkova, I., Ozbakkaloglu, T., Natin, N., Karelina, M., & Azevedo, A. (2021). Design strategy for recycled aggregate concrete: A review of status and future perspectives. *Crystals, 11*(6), 695. https://doi.org/10.3390/cryst11060695

Medina, C., Rojas, M. I. S. D., & Frias, M. (2013). Properties of recycled ceramic aggregate concretes: Water resistance. *Cement and Concrete Composites, 40*(1), 21–29. https://doi.org/10.1016/j.cemconcomp.2013.04.005

Meng, T., Wei, H., Yang, X., Zhang, B., Zhang, Y., & Zhang, C. (2021). Effect of mixed recycled aggregate on the mechanical strength and microstructure of concrete under different water cement ratios. *Materials, 14*(10), 2631. https://doi.org/10.3390/ma14102631

Nakhi, A. B., & Alhumoud, J. M. (2019). Effects of recycled aggregate on concrete mix and exposure to chloride. *Advances in Materials Science and Engineering, 2019*(1), 1–7. https://doi.org/10.1155/2019/7605098

Nara, E. O. B., Moraes, J. A. R., de Freitas, A. M. V., Rediske, G., & Benitez, G. B. (2014). Addition of alternative materials to ceramic slabs. *Ceramica, 60*(355), 1–10. https://doi.org/10.1590/S0366-69132014000300005

Ncube, L. K., Ude, A. U., Ogunmuyiwa, E. N., Zulkifli, R., & Beas, I. N. (2021). An overview of plastic waste generation and management in food packaging industries. *Recycling, 6*(1), 12. https://doi.org/10.3390/recycling6010012

Needhidasan, S., Samuel, M., & Chidambaram, R. (2014). Electronic waste—An emerging threat to the environment of urban India. *Journal of Environmental Health Science and Engineering, 12*(36), 1–20. https://doi.org/10.1186/2052-336X-12-36

Poulos, H. G. (2016). Tall building foundations: Design methods and applications. *Innovative Infrastructure Solutions, 1*(10), 1–51. https://doi.org/10.1007/s41062-016-0010-2

Prakash, K. S., & Rao, C. H. (2016). Study on compressive strength of quarry dust as fine aggregate in concrete. *Advances in Civil Engineering, 2016*(1), 1–5. https://doi.org/10.1155/2016/1742769

Senthamarai, R. M., & Monaharan, D. P. (2005). Concrete with ceramic waste aggregate. *Cement and Concrete Composites, 27*(9–10), 910–913. https://doi.org/10.1016/j.cemconcomp.2005.04.003

Subedi, M. M. (2013). Ceramics and its importance. *The Himalayan Physics, 4*(4), 80–82.

Tabak, Y., Kara, M., Gunay, E., Yildirim, S.T., & Yilmaz, S. (2012). Ceramic tile waste as a waste management solution for concrete. In 3rd International Conference on Industrial and Hazardous Waste Management (pp. 1–8).

Tadesse, H. T. (2020). Investigation on the appropriateness of ceramic waste materials at hawassa ceramics factory as partial replacement of conventional aggregate in concrete making. *IOP Conference Series: Materials Science and Engineering, 993*(1), 012123. https://doi.org/10.1088/1757-899X/993/1/012123

Tokittikul, P., & Chaipanich, A. (2010). Utilization of ceramic waste as fine aggregate within Portland cement and fly ash concretes. *Cement and Concrete Composites, 32*(6), 440–449. https://doi.org/10.1016/j.cemconcomp.2010.02.004

Troschinetz, A. M. (2005). Twelve factors influencing sustainable recycling of municipal solid waste in developing countries solid waste in developing countries. Master Thesis, Michigan Technological University (pp. 1–139). https://digitalcommons.mtu.edu/cgi/viewcontent.cgi?article=1276&context=etds

Vieira, C. M. F., & Monteiro, S. N. (2009). Incorporation of solid wastes in red ceramics: An updated review. *Materia (rio De Janeiro), 14*(3), 1–10. https://doi.org/10.1590/S1517-70762009000300002

Xiao, J., Wang, C., Ding, T., & Akbarnezhad, A. (2018). A recycled aggregate concrete high-rise building: Structural performance and embodied carbon footprint. *Journal of Cleaner Production, 199*(1), 868–881. https://doi.org/10.1016/j.jclepro.2018.07.210

Yehia, S., Abdelfatah, A., & Mansour, D. (2020). Effect of aggregate type and specimen configuration on concrete compressive strength. *Crystals, 10*(7), 625. https://doi.org/10.3390/cryst10070625

Zanelli, C., Conte, S., Molinari, C., Soldati, R., & Dondi, M. (2021). Waste recycling in ceramic tiles: A technological outlook. *Resources, Conservation and Recycling, 168*(1), 105289. https://doi.org/10.1016/j.resconrec.2020.105289

Zegardlo, B., Szelag, M., & Ogrodnik, P. (2018). Concrete resistant to spalling made with recycled aggregate from sanitary ceramic wastes—The effect of moisture and porosity on destructive processes occurring in fire conditions. *Construction and Building Materials, 173*(1), 58–68. https://doi.org/10.1016/j.conbuildmat.2018.04.030